Journal of Inherited Metabolic Disease

Production Editor: R publishing Company

This journal are also available postnatally priced ...
Copies ... and be sent the answer to the ...
Published by ... PO Box ...

Journal of Inherited Metabolic Disease

This review issue is also available separately, price Dfl. 130.– [ISBN 0-7923-8837-2].
Orders should be sent to: Kluwer Academic Publishers Group, PO Box 322, 3300 AH Dordrecht, The Netherlands, or at PO Box 358, Accord Station, Hingham, MA 02018-0358, USA, or to your local specialist bookseller.

J. Inher. Metab. Dis. 16 (1993) 613
© SSIEM and Kluwer Academic Publishers. Printed in the Netherlands

Preface

THE 30TH ANNUAL SYMPOSIUM OF THE SSIEM - LEUVEN, 1992

The 30th Annual Symposium of the SSIEM was held in Leuven in the Maria-Theresia college from September 8th to 11th 1992. 'Inherited Metabolic Diseases and the Brain' was the theme of this jubilee meeting. The Leuven meeting thus continued a tradition since this was also the main topic of the first (Sheffield 1963) and of other SSIEM Symposia.

The minisymposium on Tuesday, September 8th was on 'Carbohydrate-deficient glycoprotein syndromes'. A panel of speakers from Belgium, Sweden and the Netherlands covered all aspects of this rapidly expanding group of genetic diseases discovered in Leuven in 1980. At the end of the minisymposium, and in the presence of Mrs S. Komrower, Dr Brian Fowler announced the inauguration of the George Komrower Memorial Lectureship and gave a captivating overview of the life and work of Dr George Komrower. The lectureship commemorates the achievements of this pioneer in the field of inherited metabolic diseases and his loyal support to the SSIEM. The first George Komrower Memorial Lecture will be given at the 1993 symposium in Manchester.

The Main Symposium began on Wednesday, September 9th after the welcome to the Katholieke Universiteit Leuven given by the Vice-Rector, Professor Herman Van den Berghe. Plenary sessions were devoted to Inborn Errors and Brain Fluids, Neurotransmitter Disorders, Inborn Errors and Demyelination, and Recent Developments. Ten outstanding free papers were chosen for oral presentation and from 236 abstracts, 203 were chosen for poster display. The large number of posters in a restricted area imposed some difficulties. We would like to thank all the contributors and express our appreciation for their patience and understanding.

The attendance in Leuven was a record for our annual symposium with 371 scientific delegates (from 29 countries) plus 15 for single days as well as 23 social participants. We are very grateful for the enormous efforts of the local organizers, namely Baudouin François, Georges Van den Berghe, Peter Declercq, Marthe Jaeken and Nadine Kerstens. Many companies and organizations contributed financially towards the success for the meeting. In particular, The Society wishes to express its appreciation to Mr Brian Gill for the continuing support from Scientific Hospital Supplies Ltd. of Liverpool. Last but not least, thanks to the Kluwer donation and the Komrower bequest nine scholarships could be offered to participating junior scientists.

R. A. Harkness, J. Jaeken, G. M. Addison, G. T. N. Besley, R. J. Pollitt

The following paper was also presented at the meeting. A script was not available by the time of publication. Amino acid transport at the blood–brain barrier. Q. Smith, Bethesda, MD, USA.

J. Inher. Metab. Dis. 16 (1993) 614–616
© SSIEM and Kluwer Academic Publishers. Printed in the Netherlands

Introduction of the Komrower Commemorative Lecture Leuven 1992

It was my honour to announce the introduction of the George Komrower Memorial Lecture during the 1992 Annual Symposium in Leuven in the presence of Mrs Komrower. The lectureship has been founded with a gift from Mrs Komrower and a bequest from Dr Komrower whose death on the 11th December 1989 was reported in our Newsletter of spring 1990.

His many colleagues throughout the world will remember him as a scholarly, inspirational man of great charm. His highly active career as a paediatrician was acknowledged in obituaries in the national press in the U.K. and in medical publications (*Lancet* 1990; **1**: 404; *Br Med J* 1990; **300**: 324). His great contribution to the field of inborn errors of metabolism was formally recognized in 1982 when he was elected as an honorary member of our Society.

Dr Komrower was born on the 31st December 1911. He began his career in paediatrics when he joined the staff of the Royal Manchester Children's Hospital and St Mary's Hospital in Manchester at the beginning of 1948.

His first major contribution to our field was in the 1950s in collaboration with Ms Vera Wilson and Dr Victor Schwarz. In pioneering work on the elucidation of the defect in galactosaemia the intuitive use of red cells led to the identification of the metabolic block in this disorder. They showed that both *in vivo* and *in vitro* excessive levels of galactose led to the accumulation of galactose-1-phosphate (Schwarz *et al* 1956). This pointed to gal-1-P uridyl transferase deficiency as the cause of this condition. Later he successfully treated his patients with a low galactose diet overcoming many practical difficulties.

After his retirement in 1975 he held honorary positions in Oxford and Manchester and from 1979 to 1982 he was President of the British Paediatric Association. In 1981 Dr Komrower was honoured by our Society when he was invited to give the F.P. Hudson memorial lecture at the Annual SSIEM Symposium held that year in Southampton (Komrower 1982).

His article in the Proceedings of that meeting titled 'Galactosaemia – Thirty years on. The experience of a generation' relates his personal view of his early research on this disorder and illustrates his ability to identify remaining pathogenetic and management problems. Particularly impressive is his foresight in drawing our attention to the rather poor outcome of patients thought to have been adequately treated.

His involvement in the study of homocystinuria began in 1963, soon after the first report of this condition when he reported the first case identified in the Royal Manchester Children's Hospital (Komrower and Wilson 1963). By 1967 he had identified further cases as well as performing basic research particularly in collaboration with Dr Paul Wong from Chicago and Prof. Charles Dent's group in London.

His pioneering role in newborn screening began at about the same time when he was one of a number of workers who recognized the need for early diagnosis and introduction of treatment. In 1967 he introduced an initial study for newborn screening for phenylketonuria in the Manchester area. He had the foresight to use the method of one-dimensional paper chromatography which had the advantage of also detecting other amino acid disorders including homocystinuria due to cystathionine synthase deficiency (Komrower et al 1968).

At the 1970 meeting of our Society held in Belfast on Inherited Disorders of Sulphur metabolism he presented his experience of the dietary treatment of nine children with homocystinuria (Komrower and Sardharwalla 1971). The theoretical basis of treatment was critically considered and the central aim of reducing homocystine levels using a low methionine diet or pyridoxine or betaine was clearly argued. The question of whether to modulate other metabolites such as cystathionine or cystine was rationally considered. Also the practical difficulties relating to the dietary preparations were emphasized as well as the importance of home and social conditions in the long-term management of patients on such special diets.

As well as his pioneering efforts as a paediatrician and researcher Dr Komrower exhibited great skills as an organizer. He applied vision, determination and enthusiasm to reach the goal of providing the best possible care for his patients. This resulted in the establishment of a diagnostic and research unit in Manchester which is dedicated solely to inherited metabolic disorders. This began as the Mental Retardation Research Unit in 1961 in a modified kitchen in the Royal Manchester Children's Hospital with a part-time biochemist and one technician all financed by research grants. During the following decade the laboratory and clinical services were greatly expanded. The Unit's national and international reputation and the quality of its work helped Dr Komrower to convince the local state health authority to take over the costs of its services thereby providing a firm financial basis for the screening, diagnosis and care of patients with inherited metabolic disease.

Following further expansion and development of its services under the direction of his successor, Dr I.B. Sardharwalla, the unit moved at the end of 1984 into a new purpose-built establishment. Today the so-called Willink Biochemical Genetics Unit provides a comprehensive clinical and laboratory service on a supra-regional basis for a wide range of inherited metabolic diseases.

This unit is a testament to George Komrower's enormous personal efforts and to his inspiration of those of us who were fortunate to work with him and to continue the activities which he initiated 40 years ago. From my personal contact with him over 25 years I can vouch for his great kindness in supporting and guiding his students and younger colleagues and for his great ability to stimulate others by example and discussion, qualities which he exhibited right up to the end of his life.

It is very appropriate that the first George Komrower Memorial Lecture will be given at the next SSIEM meeting in his home town of Manchester and that the subject will be 'The Molecular Basis of Phenotype Expression in Homocystinuria due to CS deficiency'. This will be given by Prof. Jan Kraus from Denver.

I am sure that this and future memorial lectures will underline one of the most important and fundamental philosophies upheld by our Society and so ably practised by Dr Komrower. That is the furtherance of knowledge by applying basic scientific investigation to clinical problems with the overall aim of improving the care of our patients.

Dr Brian Fowler – Basel, March 1993

REFERENCES

Komrower G.M. (1982) Galactosemia – Thirty years on. The experience of a generation. *J Inher Metab Dis* **5**: 96–104.
Komrower GM, Fowler B, Griffiths MJ, Lambert AM (1968) A prospective community survey for aminoacidaemias. *Proc R Soc Med* **61**: 294–296
Komrower GM, Sardharwalla IB (1971) The dietary treatment of homocystinuria. In Carson NA, Raine DN, Inherited Disorders of Sulphur Metabolism, Proceedings for the Eighth Symposium for the Society for The Study of Inborn Errors of Metabolism. Edinburgh and London: Churchill Livingstone, 254–263
Komrower GM, Wilson VK (1963) Homocystinuria. *Proc R Soc Med* **56**: 26–27
Schwarz V, Goldberg L, Komrower GM, Holzel A (1956) Some disturbances of erythrocyte metabolism in galactosaemia. *Biochem J* **62**: 34–40.

J. Inher. Metab. Dis. 16 (1993) 617–638
© SSIEM and Kluwer Academic Publishers. Printed in the Netherlands

Extracellular and Cerebrospinal Fluids

M. B. SEGAL
Sherrington School of Physiology, United Medical and Dental Schools, St. Thomas's Hospital, London, SE1 7EH, UK

Summary: The mechanism of formation of extracellular fluid is first described, followed by an explanation of the relation between osmotic force, reflection coefficient and molecular size. The possible mechanism of brain extracellular fluid formation is then proposed in relation to the restriction offered by the blood–brain barrier.

   The functions and compositions of cerebrospinal fluid (CSF) are then described followed by sections on the process of formation of CSF, the non-electrolytes and proteins in CSF, the drainage mechanisms and protein synthesis by the choroid plexus.

EXTRACELLULAR FLUID

In the majority of vascular beds within the circulation the capillaries are freely permeable to all molecules with a molecular weight of less than 10 000 daltons. These vessels are the site of production of the extracellular fluid (ECF), also called the interstitial fluid. The routes by which fluid and small molecules pass across the capillary wall are between the cells of the endothelium, through the fenestrations in the capillary wall, present in some vascular beds, and directly through the endothelial cell walls. The histological dimensions of these routes, apart from that across the cell, do not present a barrier to even large molecules within the plasma, and the selectivity of these pathways to macromolecules resides in the structure of basement membrane, which lies beneath the endothelium. The matrices of this fibrillar basement membrane, a loose negatively charged meshwork, permits the free passage of all small molecules, but retains the plasma proteins and molecules above 60 000 molecular weight (Michel 1985). However, some plasma proteins do pass into the ECF, probably the venular side of the microcirculation, and must be returned to the blood, which is one of the prime functions of the lymphatic system.

   The rate of formation of ECF from a capillary depends on the balance between the hydrostatic pressure that forces fluid out of the vessel and the colloid osmotic pressure (COP) generated by the concentration of the plasma proteins in plasma, which opposes this movement. This latter force, also called the oncotic pressure, has an effective value of 25–28 mmHg. As can be seen in Figure 1a, fluid and small molecules move out of the arteriolar end of the capillary, since the hydrostatic

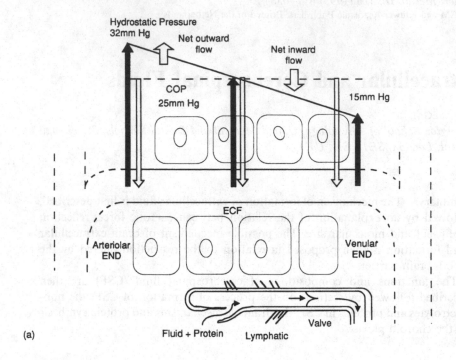

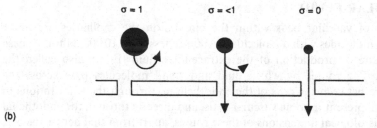

Figure 1 (a) Diagram showing filtration and absorption of fluid by the capillary endothelium. At the arteriolar end of the capillary, the hydrostatic pressure (HP) is 32 mmHg, which drives fluid out of the vessel. This movement is opposed by the colloid osmotic pressure (COP) of the plasma proteins, which has a value of 25 mmHg, so that there is a net movement of fluid out of the vessel in this region. However, as the HP falls along the length of the capillary, the forces balance by the mid-region and then the COP exceeds the HP, leading to a net uptake of fluid. Some leakage of protein does occur, which would lead to a decrease in the COP gradient, but this protein and fluid is recovered by the lymphatics. (b) The reflection coefficient σ refers to the dimensions of the pathways through membranes in relation to the size of molecules under consideration. If dimensions of the molecule are greater than the pathways, σ = 1 and the full osmotic pressure of that molecule will be developed across the membrane. If molecules can freely permeate the membrane because its dimensions are much less than the pathway, σ = 0 and no osmotic force from the molecule will develop. Intermediate values of σ indicate limited permeation of a molecule, so the osmotic pressure developed will be less than predicted from the molar concentration of the molecule

pressure of about 32 mmHg exceeds that of the COP, whereas at the venular end the hydrostatic pressure has now fallen to 12–15 mmHg, so fluid now moves back into the capillary as the COP now exceeds the hydrostatic pressure. This balance is termed the *Starling equilibrium* and may be written formally as

$$J_v = L_p S(P_c - P_e) - \sigma(\Pi_p - \Pi_e)$$

J_v is the flow; L_p is the hydraulic conductivity, which relates to the degree of permeability of the wall and S the surface area; P_c is the hydrostatic pressure driving fluid out of the lumen and P_e the pressure in the ECF, which is negative in all vascular beds except the heart. Π_p is the COP of plasma and Π_e that of the ECF, which is normally much less than that of plasma. σ is the reflection coefficient and in this case indicates the degree of 'leakiness' of the capillary wall to plasma proteins; if there is no leak it has the value of 1 and values less than this if some of the macromolecules escape across the capillary wall, which reduces their effective osmotic pressure. In the gut σ is less than 1 and plasma proteins are found in the intestinal lymph at concentration near to that in plasma, whereas in muscle σ is much closer to 1 and the lymph has a low concentration of protein. Oedema is an imbalance of this mechanism and may be caused by such factors as raised permeability to protein, low plasma protein or raised venular pressure, etc. (Levick 1991).

The osmotic force developed by a molecular species in a biological system depends on the number of 'particles' in the solution and the reflection coefficient of the membrane to these molecules across which the osmotic gradient has to be established:

$$\Pi = \sigma RTC$$

where R is the gas constant, T the temperature in K and C the molar concentration. For the proteins in plasma at a concentration of 50–80 g/L, the osmotic pressure developed is quite small, around 25 mmHg, since although the value of σ is close to 1 the number of these large molecules at given concentration is relatively 'low'. In contrast, if we consider sodium chloride, with a low molecular weight, the number of 'molecules' at a concentration in plasma of 150 mmol/L (9 g/L) is vast and since NaCl dissociates fully this number is doubled. In theory if an epithelium increases the concentration of NaCl by 1.5 mmol/L on the side relative to the other, the potential pressure developed could be 60 mmHg! However, such large potential osmotic pressures can only be generated if the σ of the membrane for NaCl is close to unity and this value is only approached in very tight epithelia (Figure 1b). In contrast, in a capillary the σ for NaCl is close to zero, so no osmotic force will be generated by the electrolytes in plasma.

From the preceding paragraphs, it can be seen that to produce an extracellular fluid by the passive Starling mechanism, the capillary wall must be freely permeable to all small molecules or large ionic osmotic pressures will be generated, which will oppose such movement.

BRAIN EXTRACELLULAR FLUID

The brain and spinal cord, being enclosed within a rigid bony case, cannot expand in the event of a disturbance in fluid balance. Some expansion can occur by the expulsion of cerebrospinal fluid (CSF) from the system, but this is very limited. The brain is thus faced with a need to control strictly the entry of fluids across the cerebral capillaries and it is to this end that the blood–brain barrier has evolved. By the end of the nineteenth century it was well recognized that a restriction was present between blood and brain by Erhlich and colleagues; the classic experiments of Goldman placed the concept of the blood–brain barrier on a firm experimental foundation, see Bradbury (1979) for a review. The site of this barrier however, remained a matter of considerable debate, even though Rodriguez (1955), with his fluorescent dye studies, clearly demonstrated that the restriction must reside at the level of cerebral endothelium; this was later confirmed by the elegant electron microscopy of Reese and Karnovsky (1967).

Another matter of debate was whether the brain had a normal extracellular space, since the early electron microscopists had observed that the gap between neurons was only some 200 Å, which gave an approximate ECF volume of only 5–10% (Wyckoff and Young 1956). At about this time the whole concept of a blood–brain barrier was challenged by others who postulated that the blood–brain barrier was a histological artefact. These ideas however, were soon disproved by Davson and colleagues, using the rabbit, who showed that when marker molecules were kept at constant levels in the blood, rapid equilibration occurred into muscle but not into brain. However, when the brain was sliced thinly and incubated *in vitro*, these same molecules now penetrated into the brain at a rate equal to that into diaphragmatic muscle incubated at the same time (Davson and Spaziani 1969) (Figure 2). From these experiments it was clearly demonstrated that the brain had an extracellular space of about 15%, which is about the same size as that of other tissues.

In vivo it is difficult to measure the extracellular space because the blood–brain barrier restricts the entry of marker molecules. Any marker molecules that do cross the blood–brain barrier and enter the brain ECF will diffuse into the CSF — the sink action of CSF — and be returned by the CSF drainage mechanisms to the blood, so keeping the brain value low (Davson et al 1987). If we assume that the ECF of the brain is some 15% of the brain weight, the volume of this fluid in the human will be in the region of 250–300 ml, which is approximately twice the volume of the CSF. From the preceding discussion on the formation of ECF, it is obvious that brain ECF fluid cannot be formed by a simple passive Starling mechanism, since the blood–brain barrier restricts the movement of all small polar molecules and moving water, without its accompanying electrolytes, would generate a large opposing osmotic force. The first person to recognize this was Davson, who in 1956 wrote in his *Physiology of the Ocular and Cerebrospinal Fluids*, 'We cannot exclude the possibility that the fluid (extracellular) is secreted by the cells of the capillary endothelium, however revolutionary such a hypothesis may seem; if this were true, the rate of formation would be rigorously controlled; a defect in the secretory system — for example, the development of numerous and large leaks in the capillaries — would circumvent this control, permitting a much more rapid loss of fluid into the nervous tissue than normally occurs, and this would give rise to the phenomenon of oedema.'

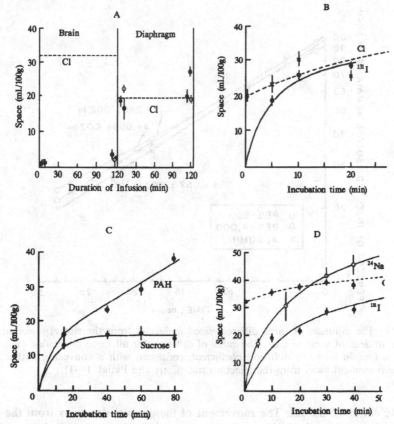

Figure 2 (A) The penetration of PAH ($\otimes$), sucrose ($\bullet$) and ^{131}I ($\bigcirc$) into the brain and diaphragmatic muscle in the rabbit *in vivo*. The marker molecules were infused for 6 or 120 min. On the left-hand side is shown the limited entry of these molecules into brain, whereas on the right-hand side is shown the rapid equilibration into the extracellular space of the muscle of the diaphragm. (B, C and D) Same tissues incubated *in vitro* as strips of muscle (B) and slices of brain (C and D). Note that now the marker molecules penetrate at the same rate into both tissues since the blood–brain barrier is now bypassed and that the brain has an extracellular space of the same magnitude as that in muscle. (Davson and Spaziani 1959)

MECHANISM OF FORMATION OF BRAIN ECF

To investigate whether the brain ECF was secreted by the blood–brain barrier the effect of various sodium transport inhibitors on the rate of entry of isotopic sodium from blood into brain was investigated (Davson and Segal 1970). These experiments failed to demonstrate any difference in the entry of sodium into the brain between the test and control animals, which may reflect the insensitivity of the method used. However, using an alternative approach, the turnover of brain ECF has been demonstrated by Cserr and colleagues (1991). The initial experiments studied the movement of a series of molecules ranging in size from small polythene glycols (900 and 4000 daltons) to albumin (69 kDa), which were microinjected, in small volumes,

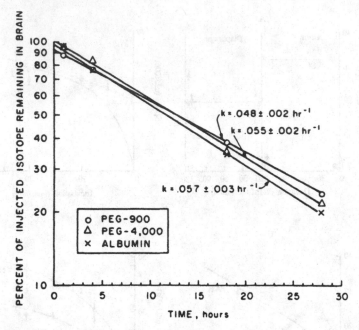

Figure 3 The diffusion of three different-sized molecules from the site of injection in the caudate nucleus of a rat *in vivo*. The rates of diffusion for all three molecules were similar despite a fivefold range in diffusion coefficient, consistent with a convective rather than a diffusive movement away from the injection site. (Cserr and Patlak 1991)

into the cortex of the rat. The movement of these molecules away from the site of injection was followed and it was found that the movement for all three molecules fitted a single rate constant, even though their diffusion coefficients differed by a factor of 5 (Figure 3). This must indicate that the fluid is moving by convective bulk flow and not by simple diffusion (see Cserr and Patlak 1991). From these data there can be no doubt that there is a slow current of brain ECF moving through the brain into the CSF, which must reflect the active secretion of this fluid by the cells of the cerebral endothelium. The rate of production of brain ECF could be calculated from these results; for the rat the value was $0.2\,\mu l/g$ per min and for the rabbit $0.11\,\mu l/g$ per min. These values are about one-tenth of the rate of formation of CSF, which would suggest the brain ECF is a rather stagnant fluid with a much lower rate of turnover than that of CSF.

Further support for the role of cerebral endothelium in the secretion of brain ECF comes from the observations of Oldendorf and colleagues (1977) who found that cerebral endothelial cells contain mitochondria that are rarely seen in other vascular beds. Na/K ATPase has also been located in the albuminal side of these cells, which again is evidence that active sodium transport is occurring in this tissue (Betz et al 1980).

The importance of regulation of brain ECF volume has lead Cserr and colleagues to study the effect of acute changes in plasma osmolarity induced with NaCl or with

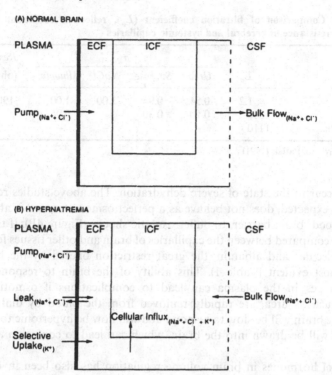

Figure 4 A model of fluid movements within the brain. The blood–brain barrier is indicated by the heavy black line separating the plasma from the extracellular fluid (ECF) of the brain. The inner square is the intracellular fluid (ICF) of the brain cells and the dashed line represents the permeable ependyma separating the brain ECF from the CSF. (A) In the normal brain the ECF is formed by the active transport of NaCl across the blood–brain barrier. The ECF formed drains into the CSF by convective bulk flow. (B) During hypernatraemia, the cells of the brain are protected from excessive shrinkage by the uptake of Na^+, Cl^-, and K^+ from both the CSF and plasma, the ICF volume being conserved by the influx of ions from the brain ICF. In this state the normal flow of brain ECF towards the CSF is reversed. (Cserr and Patlak 1991)

mannitol (Cserr and Patlak 1991). The response to plasma hyperosmolarity was to reverse the normal direction of flow of brain ECF, from brain to CSF (Figure 4A), so now sodium chloride and water were taken up into the brain from the CSF; this, however, only accounted for about 70% of the response and the rest was by movement of fluid across the blood–brain barrier, as a consequence of the osmotic gradient and also by active transport (Figure 4B). For potassium there was a difference and it seems that for this ion there is a selective uptake by the blood–brain barrier. With longer-term changes in osmolarity the brain is able to maintain the volume of neurones and other cells in the CNS, which is thought to be achieved by the generation of ideogenic osmolytes such as amino acids, for example taurine (Lohr et al 1988; Huxtable 1992). These responses enable the brain cells to maintain their cell volume and presumably protect brain function from the effect of these fluid shifts,

Table 1 Comparison of filtration coefficient (L_p), reflection coefficient (σ) and electrical resistance of cerebral and systemic capillaries

Capillary	L_p	σ				Resistance (ohm cm^2)
		Urea	Sucrose	NaCl	Albumin	
Cerebral	1.2	0.54	0.96	1.00	1.00	1900
Heart	310	0.10	0.30	–	0.96	–
Mesenteric	1110	–	–	–	–	1–3

From Cserr and Patlak (1991)

which can occur in the state of severe dehydration. The above studies revealed that the brain, as expected, does not behave as a perfect osmometer, and is able, with the aid of the blood–brain barrier, to buffer osmotic shocks (Figure 4B). If the values of L_p and σ are compared between the capillaries of brain and other tissues for a number of small molecules and albumin, the great restriction offered by the blood–brain barrier is most evident (Table 1). This ability of the brain to respond slowly to osmotic changes in the plasma can lead to complications if osmotically active molecules, such as urea, are rapidly removed from the blood by dialysis. In this condition the brain will be slow to respond and will now be hypertonic to the plasma, and so fluid will be drawn into the brain, which can lead to confusion and cerebral oedema.

The role of hormones in brain volume regulation has also been investigated by Cserr and colleagues using the ADH-deficient Brattlebro Rat. Interestingly this strain of rat was unable to respond to hyperosmotic stress as well as normal animal, which points to a central action of vasopressin and other peptides in the regulation of brain volume (De Pasquale et al 1989).

THE COMPOSITION OF BRAIN ECF

In other tissues the composition of lymph draining from a tissue can be taken as equivalent to the ECF of that tissue. However, the same is not true for the ECF of brain and CSF since the two fluids are formed by very different tissues, the cerebral endothelium and the choroid plexuses, respectively. In addition, the brain ECF may also be modified by exchanges with the neurones and glia. For the direct measurement of the composition of brain extracellular space, ion-selective electrodes have been used, and for potassium most values agree with values obtained by equilibration studies using ventriculo-cisternal perfusion (Bradbury and Davson 1965; Heinemann and Lux 1977). However, some studies point to an even lower value of 2.2 mmol/L (Cole and Poulain 1990; Cserr et al 1991).

Undoubtedly the most exciting method for the study of brain ECF has been Ungerstedt's microdialysis technique (Ungerstedt 1984). This method uses a small section of renal dialysis tubing, sealed at one end and attached to a double-lumen tube (Figure 5) (Benveniste 1989). With the aid of a stereotactic frame the probe can be placed into specific brain regions, then cemented in place, and the animal allowed to recover. The ECF can then be sampled in the conscious freely moving animal and

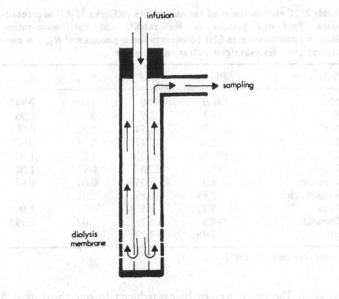

Figure 5 A section through a dialysis probe. The fluid is introduced down the central tube and passes out in the region of the dialysis membrane. The flow is slow and the fluid equilibrates with the brain ECF through the dialysis membrane. The fluid exits via the outer tubing. (Benveniste 1989)

the effect of physiological stimuli on the release of neurotransmitters such as acetylcholine, norepinephrine, dopamine, etc., can be evaluated in specific brain regions. The effect of drugs and their actual local concentration in the brain can also be measured, so this technique can be applied to a wide range of problems. (The company CMA/Microdialysis of Sweden produces a complete bibliography of all papers using this technique.)

In conclusion, brain ECF has on theoretical grounds to be produced by active transport processes at the blood–brain barrier, the mechanism of which awaits further investigation. The rate of secretion of this fluid is slow and the normal direction of flow is from the brain ECF towards the CSF. Complex homeostatic mechanisms exist to control the composition of this fluid with regard to electrolytes, neurotransmitters and amino acids; many of these latter molecules have powerful neurotransmitter and neuromodulatory actions.

CEREBROSPINAL FLUID

Cerebrospinal fluid (CSF) fills the ventricles and subarachnoid space of the brain and spinal cord. In man the volume is approximately 140 ml and it is secreted at a rate of about 0.5 ml/min. The turnover rate of this fluid is high, about 4–5 times per day. The pressure within the ventricular system is 150–180 mmH$_2$O and pressures above 200 mmH$_2$O are considered to be abnormal.

The CSF functions to support the brain and prevents damage by mechanical

Table 2 Concentrations of various solutes (mEq/kg H_2O) in cerebro-spinal fluid and plasma of the rabbit, and distribution-ratios. R_{CSF} = concentration in CSF/concentration in plasma and R_{Dial} = concentration in dialysate/concentration in plasma

Substance	Plasma	CSF	R_{CSF}	R_{Dial}
Na	148.0	149.0	1.005	0.945
K	4.3	2.9	0.675	0.96
Mg	2.02	1.74	0.92	0.80
Ca	5.60	2.47	0.45	0.65
Cl	106.0	130.0	1.23	1.04
HCO_3	25.0	22.0	0.92	1.04
Glucose	8.3	5.35	0.64	0.97
Amino acids	2.84	0.89	0.31	–
Urea	8.35	6.5	0.78	1.00
Osmolality	298.5	305.2	1.02	0.995
pH	7.46	7.27	–	–

From Davson et al (1987)

insults to the skull; the brain weight being reduced to one-third as it floats in the CSF. The CSF may act as a drainage pathway for waste products, electrolytes and excess neurotransmitters since the brain does not have a lymphatic system. This function, termed the 'sink action' of the CSF by Davson, is aided by the low level of many non-electrolytes in CSF and the flow of brain ECF from the brain towards the ventricular and subarachnoid compartment (Davson et al 1987; Cserr and Patlak 1991). A further more speculative function of the CSF is that it may act as a 'third' circulation within the brain, carrying peptides secreted in one brain region to another, such as between the circumventricular organs (Nilsson et al 1992).

CSF has an electrolyte composition similar to plasma with a low protein concentration of 25 mg/100 ml. However, careful examination of the concentration ratio of ions in plasma and CSF, sampled at the same time, reveals small but consistent differences between the two fluids. The ratios of most ions between CSF and plasma are different from those of a dialysate of plasma, which implied that CSF was produced by a process of active secretion (Davson et al 1987) (Table 2).

SITES FOR PRODUCTION OF CSF

CSF is secreted by the choroid plexuses of the lateral, IIIrd and IVth ventricle with a small additional production of fluid from the brain ECF. The choroid plexuses are leaf-like structures that float in the CSF with a central core of blood vessels covered on both sides by an epithelium. The choroidal cells are joined together on the CSF side by an occluding band of tight junctions, the site of the blood–CSF barrier. The choroidal capillaries are of the fenestrated type and are freely permeable to all small molecules, unlike the majority of the cerebral vasculature. The epithelial cells of the choroid plexus have the typical histology of a transporting epithelium, with many mitochondria within the cytoplasm, microvilli and cilia on the CSF side, and complex intracellular clefts on the blood side of the cell (Figure 6). Freeze-fracture studies of

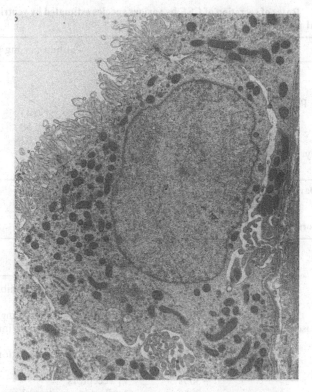

Figure 6 An electronmicrograph of a cell from the rabbit choroid plexus. On the upper left is the cerebrospinal fluid, and on this side of the cell there are prominent microvilli with tight junctions between the cells. The cell cytoplasm contains many mitochondria. On the basal (blood) side of the cell and the lateral intercellular spaces there are prominent interdigitations between the cells. The space between the cells is larger than normal as the secretion of CSF was inhibited in this animal by ouabain in the CSF. (Segal and Burgess 1974)

the choroidal tight junctions show them to be of the 'leaky' type with 6–8 strands and open pathways between the strands on the two faces of the cell (Van Deurs and Koehler 1979).

METHODS TO STUDY CSF SECRETION

The rate of formation of CSF can be measured *in vivo* by the technique of ventriculo-cisternal (VC) perfusion (Pappenheimer et al 1962). Briefly, a mock CSF containing a high-molecular-weight marker molecule that cannot escape from the system is perfused through the ventricles via needles inserted through holes drilled in the skull. The perfusate is collected from a needle placed in the cisterna magna and the rate of CSF secretion is calculated from the dilution of the marker molecule. Using the VC technique the rate of CSF secretion has been measured in a wide variety of animals including man (Table 3), and the percentage turnover and the rate of secretion of CSF per mg of choroid plexus is moderately constant across species, suggesting a

Table 3   Rates of secretion of CSF by various species estimated by ventriculo-cisternal perfusion

Species	$\mu l/min$	(% per min)	$\mu l/min$ per mg CP
Mouse	0.325	0.89	–
Rat	3.0	1.02	–
Guinea pig	3.5	0.875	–
Rabbit	10.0	0.43	0.43
Cat	20.0	0.45	0.50
Dog	50.0	0.40	0.625
Monkey	41	–	–
Man	350.0	0.38	–

From Davson et al (1987)

Table 4   Inhibitors of CSF secretion

Inhibitor	Effect on CSF secretion	Possible mechanism
Acetazolamide	−70%	Carbonic anhydrase inhibitor
Ouabain	−70%	Na/K ATPase inhibitor
Amiloride	−50%	Na uptake Na/H exchange
Bumetanide, furosemide	−45%	Na/K/2Cl Cotransport inhibitors
Omeprazole	−35%	H/K ATPase inhibitor
DNP	−50%	Oxidative phosphorylation
VIP	−30%	cAMP?
ADH	−50%	V_1 receptors
ANP	−70%	Amiloride-sensitive channels
Glucocorticoids	−30%	
Carbamyl chloride	−20%	Cholinergic agonist
5HT (high concentration)	−30%	$5HT_{1c}$?
Noradrenaline	−30%	β-Receptors, cAMP
Angiotensin II	± ?	
Cholera toxin	+30%	cAMP

common underlying secretory mechanism. The effects of a wide variety of transport inhibitors have been tested on CSF secretion in order to determine the basic mechanism of secretion and to find a drug that will inhibit the secretion of CSF yet is sufficiently non-toxic for use in patients with obstructed CSF outflow (Table 4).

Although the *in vivo* VC technique is simple, the data are always complicated by the exchanges that may occur across the permeable ependyma lining the ventricles between mock CSF, brain ECF and the blood–brain barrier. To separate these two systems, various *in vivo* and *in vitro* preparations have been developed. Ames and colleagues (1965) exposed the choroid plexus of the cat's lateral ventricle *in vivo* and covered the plexuses with oil, so that the drops of the newly formed CSF could be collected as they were secreted into the oil. Special micro methods were then used to analyse the composition of the newly secreted drops of CSF. The composition of this fluid was similar to bulk-phase CSF and revealed that the K^+ concentration in

CSF is kept very constant in spite of wide variations in the plasma level of this ion. The choroid plexus of the lateral ventricles can be removed and survive isolation *in vitro* in a suitable oxygenated medium. This technique has revealed that the choroid plexus can accumulate many molecules from the CSF, giving high tissue/medium ratios. *In vivo* the blood flow will remove such products from the choroidal ECF and such gradients do not normally occur, so data from *in vitro* findings must be accepted with caution (Deane and Segal 1978). An alternative technique has been to mount the single-sided choroid plexus from the IVth ventricle of the bull frog in a miniature Ussing chamber (Wright 1972). This is a most useful preparation for some studies but only secretes CSF slowly and does not show a net transport of amino acids. To overcome these problems, Pollay and colleagues (1972) developed the isolated perfused choroid plexus of sheep, which secretes CSF at close to a normal rate and can be used to study both 'sides' of the choroidal epithelium.

MECHANISMS OF CSF SECRETION

The mechanism of secretion of CSF appears to possess many of the features found in most transporting epithelia and can be inhibited by a variety of agents (Table 4). One of the first compounds found to inhibit CSF secretion was acetazolamide (Diamox), the carbonic anhydrase inhibitor, which has been extensively studied by Maren and colleagues (Maren and Broder 1976). There is no doubt that carbonic anhydrase is involved in the secretion process and that the hydration of carbon dioxide has a central role. Ouabain, the potent inhibitor of Na/K ATPase, blocks CSF secretion by some 70% and this enzyme has been localized on the apical (CSF) side of the epithelium by both antibodies to the enzyme and ouabain binding studies (Ernst et al 1986; Quinton et al 1973). The choroid plexus takes up sodium from the blood side and the rate of entry of this ion into the CSF reflects the rate of CSF secretion (Davson and Segal 1970; Knuckey et al 1991). Anything that blocks Na uptake also inhibits CSF secretion, so it would appear that the water phase of this secretion may be secondary to the active transport of sodium. In addition, the replacement of Na in either the blood or the CSF blocks CSF secretion (Davson and Segal 1970).

Various groups have studied the nature of uptake processes on both sides of the choroid plexus using both *in vivo* and *in vitro* techniques and these findings point to a co-transport of Na, K and Cl on the blood side that was blocked by frusemide and bumetanide, and a Na/H exchange on the CSF side blocked both by amiloride and by omeprazole (Johanson et al 1990; Johanson and Murphy 1990; Bairamian et al 1991; Lindvall-Axelsson et al 1992). These findings and others have lead to a tentative model of the mechanism of CSF secretion by the choroid plexus and this is shown in Figure 7a from Saito and Wright (1983).

Recent studies using the powerful tool of the patch clamp technique have demonstrated chloride channels in the apical (CSF) membrane side of the choroid plexus, which would suggest that this anion is dominant in the mechanism of CSF secretion. These studies have demonstrated a Cl^-/HCO_3^- exchange carrier on the basal (blood) side of the cell that is sensitive to DIDs (4,4-diisothiocyanatostilbene-2,2-disulphonic acid) and solutions containing Cl^- and HCO_3^- lead to Cl^-

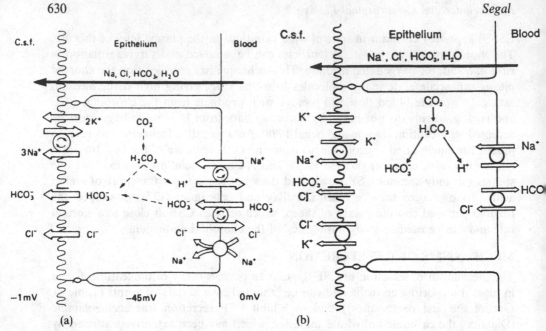

Figure 7 Models of the mechanism of CSF secretion. (a) ATPase staining and ouabain binding have identified sodium pumps on the CSF side of the choroidal cells that will extract Na^+ from the cell. On the blood side a Na/Cl co-transport aids the entry of these ions, which is inhibited by bumetanide and frusemide. Recent studies include K^+ in this process. Within the cell HCO_3^- is generated, which may pass into the CSF via HCO_3^- channels that are cAMP dependent. The process is inhibited by acetazolamide and the H^+ ions formed at the same time are exchanged for Na^+ on the blood side of the cell. (Saito and Wright 1983). (b) More recent studies using patch clamp techniques have revealed a Cl^- channel on the CSF side of the cell and a Cl^-/HCO_3^- exchange on the blood side. This would make Cl^- the most important ion in the formation of CSF. (Christensen et al 1989; Garner and Brown 1992). *In vivo* studies would favour a combination of these models, with the ionic movement generating an osmotic gradient drawing water both through the cell and through the tight junctions

accumulation in the cell as the exchange process is stimulated. The apical Cl^- channel has both a HCO_3^- as well as a Cl^- permeability and these findings have lead to a modified model shown in Figure 7b (Garner and Brown 1992). From these studies the formation of CSF depends on the active transport of sodium to generate a gradient for this ion across the choroidal cell wall and the exchange of HCO_3^- and Cl^- ions on the blood side of the cell leading to a raised intracellular concentration of the Cl^- anion in the cell. The exit of this anion via Cl^- channels in the apical surface is in some way coupled to the osmotic movement of water. A model of absorbing epithelia by Diamond and Bossert (1967) located the site of this coupling to the dilated intercellular clefts between the cells and the establishment of a standing osmotic gradient. Much criticism of this model has been voiced, but reductions in the width of these clefts are observed when transport is inhibited (Hill 1975).

Studies on the *in vivo* rabbit choroid plexus revealed that in the normal state there were no dilatations in the clefts between the cells, but when the transport of CSF was inhibited by ouabain or acetozolamide marked dilatations were seen (Segal and Burgess 1974) (Figure 6). This finding is the reverse of that seen in absorbing epithelia and may suggest that inhibition of part of the transport processes in the choroidal cell leads to cell shrinkage as ions are lost (Segal and Pollay 1977). However, the problem still remains of how the movement of ions on the CSF side of the cell is coupled to osmotic movement of water. Any ionic gradient established in the CSF close to the cell wall would rapidly be dissipated by diffusion into the bulk of CSF before it could act osmotically to draw fluid either across the cell wall or through the tight junctions.

It is of interest that in a normal healthy animal, not poisoned by an excess of the inhibitor, the CSF secretion can only be inhibited by a maximum of 70%. The drainage of brain ECF into CSF accounts for some 10% of this non-inhibited fluid, but the source of the remaining percentage is a matter of some debate (Segal and Pollay 1977).

CONTROL OF CSF SECRETION

The choroid plexus contains a wide range of receptors and nerve terminals, most of which have been identified histologically, but whose function at present has not been fully evaluated. Both sympathetic and cholinergic nerve terminals have been identified, but catecholamines and nerve section have only a small effect on CSF secretion. Other receptors for vasopressin, various peptides such as ANP and for 5HT have been located on the choroid plexus, and although some effects have been reported many of the findings are at present controversial (see review by Nilsson et al 1992; Garner et al 1992; Chodobski et al 1992; Zlokovic et al 1991). This area is being actively investigated, but at present it is difficult to evaluate whether these receptors act on the mechanisms by which CSF secretion is controlled or have other functions.

NON-ELECTROLYTES AND CHOROID PLEXUS

Cerebrospinal fluid has a lower concentration of sugar and amino acids than plasma. Studies using the ventriculo-acqueductal or VC perfusion techniques have identified the sodium-dependent uptake of both sugars and amino acids from CSF (Bradbury and Brøndsted 1973; Lorenzo 1977). This evidence has been used to suggest that the choroid plexuses are responsible for the low level of these non-electrolytes in CSF. However, studies with isolated perfused sheep choroid plexuses, using steady-state methods, have shown that there is a net entry of sugars from blood to CSF and that the low level of these molecules in CSF is the consequence of the kinetics of entry, and not the sodium-dependent efflux. From these studies, knowing both the flux and the rate of CSF secretion, the concentration of sugars in newly formed CSF could be calculated and values about 50% of that in plasma were found, which are close to those observed in bulk-phase CSF (Deane and Segal 1985).

For amino acids the isolated perfused sheep choroid plexus was also used to identify the types of carrier molecules present. In these studies the single-pass indicator dilution method was used (Yudilevich and Mann 1985). With this method the 'L' system carrier was identified for large neutral amino acids and for carriers alanine,

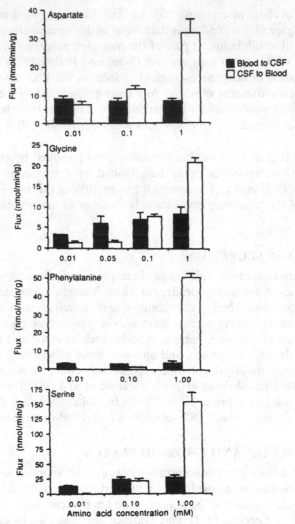

Figure 8 The steady-state amino acid fluxes across the isolated perfused sheep choroid plexus. At a low amino acid concentration there is a net flux from blood to CSF, but when the concentration is raised the flux is reversed from CSF to blood. (Preston 1989)

serine and glycine, but there was no uptake of the 'A' system analogues (Preston et al 1989). Further studies using steady-state methods revealed that, like sugars, there was a net entry of amino acids from blood to CSF and the calculated concentration of amino acid in the newly formed CSF was low and the same as that in the bulk-phase CSF. However, when the concentration of unlabelled amino acids was raised, the CSF-to-blood flux increased and there was now a net efflux from blood to CSF (Figure 8). This efflux from CSF to blood would maintain a steep gradient from brain ECF to CSF – the sink action of CSF – which may help to remove amino acids, with neurotransmitter actions, from the environment of the neurones and aid brain amino acid homeostasis (Davson et al 1987; Preston and Segal 1990, 1992a).

THYROID HORMONES AND THE CHOROID PLEXUS

Recently the choroid plexuses have been found to be a major site for the synthesis of the pre-albumin protein, transthyretin (Dickson et al 1987). This protein avidly binds thyroxine (T_4) and by the use of the isolated perfused sheep choroid plexus has shown to be secreted only in the direction from choroid plexus into CSF (Schreiber et al 1990). Earlier studies failed to demonstrate a carrier-mediated uptake of either T_4 or tri-iodothyronine (T_3) from blood into choroid plexus. However, by the use of the perfused sheep choroid plexus a relatively non-stereospecific carrier-mediated uptake from both T_3 and T_4 has been demonstrated (Preston and Segal 1992b,c). These results serve to explain the higher levels of these hormones in the *free* state in CSF with respect to plasma (Davson et al 1987). The choroid plexus/CSF route for T_4 may constitute an important component for the entry of this hormone into the brain, where it is converted to T_3, especially in the developing animal (Dratman et al 1991; Chanoine et al 1992).

CSF PROTEIN

The major difference between CSF and plasma is the very low concentration of protein in CSF, the absolute value of which is dependent on the site of sampling; lumbar CSF has a value of 40–60 mg/100 ml, whereas the value of cisternal fluid is 15–25 mg/100 ml. Age leads to a gradual increase in the protein concentration and the concentration in females is slightly less than in males. For neonates the concentrations of protein are elevated for the first 2 weeks, being 30–120 mg/100 ml, but falls rapidly after this to the normal childhood level of about 20 mg/100 ml. Electrophoresis of the CSF proteins has shown that β-lipoproteins and fibrinogen are largely absent, apart from very small traces, in normal CSF, but do appear in inflammatory conditions. Some proteins are specific to CSF, a gammaglobulin γ_c and β trace. Another group of proteins specific to CSF is a fraction migrating between the β- and γ-globulins, the τ fraction which is related to the iron-transporting proteins such as transferrin. There are also haptoglobulins and β_2 microglobulins and Table 5 shows the wide range of normal CSF proteins.

However, in general the majority of proteins in CSF reflect those that are in plasma; although the concentration of proteins in CSF is much less than in plasma, the range of proteins with respect to size is the same. This finding would imply that a leak pathway exists to proteins that is relatively non-specific, but the number of these 'large leaks' is in fact small. Figure 9 shows the relation between radius and the serum/CSF ratio (Felgenhauer 1974). The protein content of CSF is of considerable clinical interest and can be a useful tool, indicating both a breakdown of the blood–brain barrier and destruction of neural tissue. The ratio of such proteins relative to other proteins such as albumin is a useful index as the varition in normals is quite wide (Table 5). See Davson et al (1987) for a review.

Table 5 Proteins and their concentrations in serum and CSF

Protein	MW (kDa)	Serum (mg/l)	CSF (mg/l)	Serum/CSF
Prealbumin	61	238 ± 76	17.3 ± 6.6	16.2
α_1-Antitrypsin (α_1Atr)	45			228.0
α_1-Antichymotrypsin (α_1Ach)	45			216.0
Haemopexin (Hpx)	80			267.0
Albumin (Alb)	69	36 600 ± 61	155.0 ± 39.0	210.0
α_2-HS-Glycoprotein (α_2HS)	49	479 ± 116	1.7 ± 0.6	253.0
Transferrin (Tf)	81	2040 ± 260	14.4 ± 4.4	175.0
Acid α_1-glycoprotein (α_1APG)	44	664 ± 223	3.6 ± 1.4	260.0
Plasminogen (Pmg)	143	156 ± 34	0.25	662.0
Ceruloplasmin (Cp)	152	366 ± 91	0.97 ± 0.37	518.0
Immunoglobulin-G (IgG)	110	9870 ± 2200	12.3 ± 6.4	852.0
Immunoglobylin-A (IgA)	150	1750 ± 700	1.3 ± 0.6	940.0
α_2-Macroglobulin (α_2M)	798	2220 ± 650	2.0 ± 0.7	3000.0
Fibrinogen (Fbg)	840	2964 ± 639	0.65	4550.0
Immunoglobulin-M (IgM)	800	700 ± 280	0.6 ± 0.3	1166.0
β-Lipoprotein (BLP)	2239	3728 ± 709	0.59	8950.0

From Felgenhauer (1974)

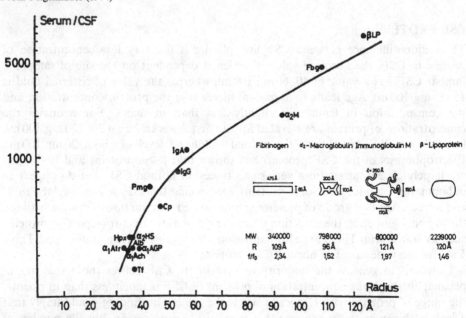

Figure 9 Correlation between serum/CSF concentration ratios and the hydrodynamic radii of plasma proteins. Inset is shown the probable shape and dimensions of four of the largest human proteins: MW = molecular weight; R = hydrodynamic radius; f/f_0 = frictional ratio; βLP = β-lipoprotein; Fbg = fibrinogen; α_2M = α_2-macroglobulin; IgA and IgG = immunoglobulins; Alb = albumin; PmG = plasminogen; Cp = ceruloplasmin; Tf = transferrin. (Felgenhauer 1974)

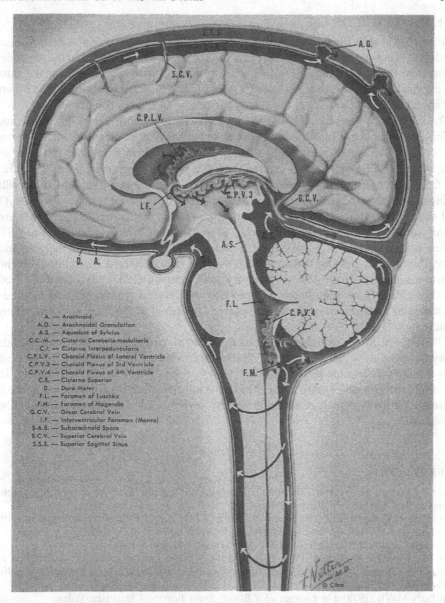

A. — Arachnoid
A.G. — Arachnoidal Granulation
A.S. — Aqueduct of Sylvius
C.C.M. — Cisterna Cerebello-Medullaris
C.I. — Cisterna Interpeduncularis
C.P.L.V. — Choroid Plexus of Lateral Ventricle
C.P.V.3 — Choroid Plexus of 3rd Ventricle
C.P.V.4 — Choroid Plexus of 4th Ventricle
C.S. — Cisterna Superior
D. — Dura Mater
F.L. — Foramen of Luschka
F.M. — Foramen of Magendie
G.C.V. — Great Cerebral Vein
I.F. — Interventricular Foramen (Monro)
S-A.S. — Subarachnoid Space
S.C.V. — Superior Cerebral Vein
S.S.S. — Superior Sagittal Sinus

Figure 10 A schematic section through the brain and spinal cord showing the drainage pathways for CSF. The CSF is formed by the choroid plexuses of the lateral ventricles (CPLV) and passes out (black arrows) by the two foramina of Munro (FM) into the third ventricle, more fluid being added by the plexus in this ventricle (CPV3). CSF then flows through the narrow aqueduct of Sylvius (AS) into the fourth ventricle, which has a single-sided sheet-like plexus (CPV4). From here fluid exits into the various basal cisterns (C) and then into the subarachnoid space (SAS) over the surface of the cortex. Some fluid drains back into the blood via the arachnoid granulation (AG) into the superior sagital sinus (SSS), some via the spinal nerve roots and the remainder via the olfactory tracts (not shown). (Netter 1953)

THE DRAINAGE OF CSF

CSF formed within the lateral ventricles drains via the two foramina of Munro into the third ventricle and then by the aqueduct of Sylvius into the fourth ventricle. From here the bulk of fluid passes by the foramina of Magendie and Luschka into the cisterna magna and then the subarachnoid space. A small quantity passes down the spinal canal to the base of the spine, but not in all species (Figure 10).

Drainage from the subarachnoid space is via the arachnoid villi or granulations, which are outpouches into the dural sinus; some are also found in the spinal nerve roots. These structures, according to the physiologists, have a valvular function so that when the CSF pressure is greater than the venous pressure, fluid drains from the CSF back to the blood. If the pressure is greater in the veins the arachnoid villi collapse until the pressure gradient is favourable. CSF pressure is 150–180 mmH$_2$O and that in the dural sinus is usually slightly subatmospheric. Adverse gradients are usually only temporary and occur during postural changes. A large fraction of CSF drains by the spinal nerve roots into the lymphatic networks of these regions, and since this fluid is passing through lymph nodes it is subjected to scrutiny by the immune system at these points. Bradbury and Westrop (1983) have demonstrated another pathway via the olfactory tracts into the cervical lymphatics that also has considerable importance in many species (see Davson et al 1987, for a review of this topic).

Obstruction of the drainage pathway for CSF by congenital malformation leads to hydrocephalus in the neonate, which can be treated by fitting a valve and tubing into the lateral ventricle so that excess CSF can then drain into the body cavity. In the adult the growth of tumours within the skull can cause obstruction of the CSF drainage pathway, which is accompanied by severe headache and elevation of the optic disc.

REFERENCES

Ames A, Higashi K, Nesbett FB (1965) The relation of potassium concentration in choroid plexus fluid to that in plasma. *J Physiol* **181**: 506–515.

Bairamian D, Johanson CE, Parmalee JT, Epstein MH (1992) Potassium co-transport with sodium and chloride in the choroid plexus. *J Neurochem* **56**: 1623–1629.

Benveniste H (1989) Brain microdialysis. *J Neurochem* **52**: 1667–1679.

Betz AL, Firth JA, Goldstein GW (1980) Polarity of the blood–brain barrier: distribution of enzymes between the luminal and abluminal membranes of brain capillary endothelial cells. *Brain Res* **192**: 17–28.

Bradbury MWB (1979) *The Concept of a Blood–Brain Barrier*. Chichester: Wiley.

Bradbury MWB, Brøndsted HE (1973) Na$^+$ dependent transport of sugars and iodide from cerebral ventricles of the rabbit. *J Physiol* **234**: 27–143.

Bradbury MWB, Davson H (1965) The transport of potassium between blood, cerebrospinal fluid and brain. *J Physiol* **181**: 151–174.

Bradbury MWB, Westrop RJ (1983) Factors influencing exit of substances from the cerebrospinal fluid into deep cervical lymph of the rabbit. *J Physiol* **339**: 519–534.

Chanoine JP, Alex A, Fang SL et al (1992) Role of transthyretin in the transport of thyroxine from blood to the choroid plexus, the cerebrospinal fluid and the brain. *Endocrinology* **130**: 933–938.

Chodobski A, Szmydynger-Chodobska J, Cooper E, McKinley MJ (1992) Atrial natriuretic peptide does not alter cerebrospinal fluid formation in the sheep. *Am J Physiol* **31**: 12860–12864.

Christensen O, Simon M, Randlev T (1989) Anion channels in a leaky epithelium: a patch clamp study of choroid plexus. *Pflug Arch* **415**: 37–46.

Cole JA, Poulain DA (1990) Local increases in extracellular [K^+] in the supra-optic nucleus during the milk ejection reflex in the anaesthetized rat. *J Physiol* **420**: 420 72P.

Cserr H, Patlak CS (1991) Regulation of brain volume under isometric and anisosmotic conditions. In Gilles R, ed. *Advances in Comparative and Environmental Physiology* 9. Berlin: Springer-Verlag, 61–80.

Cserr HF, De Pasquale M, Nicholson C, Patlak CS, Pettigrew KD, Rice ME (1991) Extracellular volume decreases which cell volume is maintained by ion uptake in rat brain during acute hypernatremia. *J Physiol* **442**: 277–295.

Davson H, Segal MB (1970) The effects of some inhibitors and accelerators of sodium transport on the turnover of ^{22}Na in the cerebrospinal fluid and the brain. *J Physiol* **209**: 131–153.

Davson H, Spaziani E (1959) The blood–brain barrier. *J Physiol* **149**: 135–143.

Davson H, Welch K, Segal MB (1987) *The Physiology and Pathophysiology of the Cerebrospinal Fluid*. Edinburgh: Churchill Livingstone.

Deane R, Segal MB (1978) Effect of vascular perfusion on the transport of sugars across the choroid plexus of the sheep. *J Physiol* **285**: 57P.

Deane R, Segal MB (1985) The transport of sugars across the perfused choroid plexus of the sheep. *J Physiol* **362**: 245–260.

De Pasquale M, Patlak CS, Cserr HF (1989) Brain ion and volume regulation during acute hypernatremia in Battlebro rats. *Am J Physiol* **256**: F1059–F1066.

Diamond JH, Bossert WH (1967) Standing gradient-osmotic flow: A mechanism for coupling of water and solute transport in epithelia. *J Gen Physiol* **50**: 2061–2083.

Dickson PW, Aldred AR, Menting JGT, Marley PD, Sawyer WH, Schreiber G (1987) Thyroxine transport in choroid plexus. *J Biol Chem* **262**: 13907–13915.

Dratman MB, Crutchfield FL, Schoenhoff MB (1991) Transport of iodothyronine from bloodstream to brain: Contributions by blood–brain and choroid plexus: cerebrospinal fluid barrier. *Brain Res* **554**: 229–236.

Ernst SA, Palacios JR, Siegel GJ (1986) Immunocytochemical localisation of Na^+, K^+ATPase catalytic polypeptide in mouse choroid plexus. *J Histochem Cytochem* **34**: 189–195.

Felgenhauer K (1974) Protein size and cerebrospinal fluid. *Klin Wochenschr* **52**: 1158–1164.

Garner C, Brown PD (1992) Two types of chloride channel in the apical membrane of the rat choroid plexus epithelial cells. *Brain Res* **591**: 137–145.

Heinemann U, Lux HD (1977) Ceiling of stimulus induced rises in extracellular potassium concentration in the cerebral cortex of cat. *Brain Res* **120**: 231–249.

Hill AE (1975) Solute-solvent coupling in epithelia: A critical examination of the standing gradient osmotic flow theory. *Proc Roy Soc B* **190**: 99–114.

Huxtable RJ (1992) The physiological actions of taurine. *Physiol Rev* **72**: 105–163.

Johanson CE, Murphy VA (1990) Acetazolemide and insulin alter choroid plexus epithelial cell [Na^+] pH and volume. *Am J Physiol* **258**: F1538–F1546.

Johanson CE, Sweeney SM, Parmalee JT, Epstein MH (1990) Co-transport of sodium and chloride by the adult mammalian choroid plexus. *Am J Physiol* **258**: C211–C216.

Knuckey NW, Fowler AG, Johanson CE, Nashold JRB, Epstein MH (1991) Cisterna magna microdialysis of ^{22}Na to evaluate ion transport and cerebrospinal fluid dynamics. *J Neurosurg* **74**: 905–971.

Levick JR (1991) *An Introduction to Cardiovascular Physiology*. London: Butterworths, 117–170.

Lindvall-Axelsson M, Nilsson C, Owman C, Winbladh B (1992) Inhibition of cerebrospinal fluid formation by omeprazole. *Exp Neurol* **115**: 394–399.

Lohr JW, McReynolds J, Grimaldi T, Acara M (1988) Effect of acute and chronic hypernatremia in myoinositol and sorbitol concentrations in the rat brain and kidney. *Life Sci* **43**: 271–276.

Lorenzo AV (⁷⁷) Factors governing the composition of the cerebrospinal fluid. *Exp Eye Res* **25**: 205- 228.

Maren TH, Broder LE (1976) The role of carbonic anhydrase in anion secretion into cerebrospinal fluid. *J Pharmacol Exp Ther* **172**: 197–202.

Michel CC (1985) The Malpighi Lecture. Vascular permeability – the consequence of Malpighi's hypothesis. *Int J Microcirc Clin Exp* **4**: 265–284.

Netter FH (1953) *The Ciba Collection of Medical Illustrations*, Vol 1. The Nervous System. Summit, New Jersey: Ciba Pharmaceutical Products Inc., 44.

Nilsson C, Lindvall-Axelsson M, Owman C (1992) Neuroendocrine regulatory mechanisms in the choroid plexus–cerebrospinal fluid system. *Brain Res Rev* **17**: 109–138.

Oldendorf WH, Cornford ME, Brown WJ (1977) Some unique ultrastructural characteristics of rat brain capillaries. *Am Neurol* **1**: 409–417.

Pappenheimer JR, Heisey SR, Jordan EF, Downer J de C (1962) Perfusion of the cerebralventricular system in unanaesthetised goats. *Am J Physiol* **203**: 763–774.

Pollay M, Stevens A, Estrada A, Kaplan R (1972) Extracorporeal perfusion of choroid plexus. *J Appl Physiol* **32**: 612–617.

Preston JE (1989) PhD thesis, University of London.

Preston JE, Segal MB (1990) The steady state amino acid fluxes across the perfused choroid plexus of the sheep. *Brain Res* **525**: 275–279.

Preston JE, Segal MB (1992a) The uptake of anionic and cationic amino acids by the isolated sheep choroid plexus. *Brain Res* **581**: 351–355.

Preston JE, Segal MB (1992b) Saturable uptake of [^{125}I]L-tri-iodothyronine at the basolateral (blood) and apical (cerebrospinal fluid) sides of the isolated perfused sheep choroid plexus. *Brain Res* **592**: 84–90.

Preston JE, Segal MB (1992c) Thyroid hormone uptake at the blood face of the isolated perfused sheep choroid plexus. *J Physiol* **446**: 87P.

Preston JE, Segal MB, Walley GJ, Zlokovic BV (1989) Neutral amino acid uptake by the isolated perfused sheep choroid plexus. *J Physiol* **408**: 31–43.

Quinton PM, Wright EM, Tarmay JMcD (1973) Localization of the sodium pump in the choroid plexus epithelium. *J Cell Biol* **58**: 724–730.

Reese TS, Karnovsky MJ (1967) Fine structural localisation of a blood–brain barrier to exogenous peroxidase. *J Cell Biol* **34**: 207–217.

Rodriguez LA (1955) Experiments on the histologic locus of the haemato-encephalic barrier. *J Comp Neurol* **102**: 27–46.

Saito Y, Wright EM (1983) Bicarbonate transport across frog choroid plexus and its control by cyclic nucleotides. *J Physiol* **336**: 635–648.

Schreiber G, Aldred AR, Jarowski A, Nilsson C, Achen MG, Segal MB (1990) Thyroxine transport from blood to brain via transthyretin synthesis in choroid plexus. *Am J Physiol* **258**: R338–R345.

Segal MB, Burgess AM (1974) A combined physiological and morphological study of the secretory process by the rabbit choroid plexus. *J Cell Sci* **14**: 339–350.

Segal MB, Pollay M (1977) The secretion of cerebrospinal fluid. *Exp Eye Res* **25**: 127–148.

Ungerstedt U (1984) Measurement of neurotransmitter release by intracranial microdialysis. In Marsden CA, ed. *Measurements of Neurotransmitter Release In Vivo*. New York: Wiley, 81–107.

Van Deurs B, Koehler JK (1979) Tight junctions in the choroid plexus epithelium. *J Cell Biol* **80**: 662–673.

Wright EM (1972) Accumulation and transport of amino acids by the frog choroid plexus. *Brain Res* **44**: 207–219.

Wyckoff RWG, Young JZ (1956) The motor-neurone surface. *Proc Roy Soc B* **144**: 440–450.

Yudilevich DL, Mann GE (1985) *Carrier-mediated Transport of Solutes from Blood to Tissue*. London: Longman.

Zlokovic BV, Segal MB, McComb JG, Hyman S, Weiss MH, Davson H (1991) Kinetics of ciculatory vasopressin uptake by choroid plexus. *Am J Physiol* **260**: F216–F224.

J. Inher. Metab. Dis. 16 (1993) 639–647

The Blood–Brain Barrier: Cellular Basis

R. C. JANZER
Institut Universitaire de Pathologie, Division de Neuropathologie, Rue du Bugon 25-27, CH-1011 Lausanne, Switzerland

Summary: Perfusion experiments with horseradish peroxidase have established that the morphological substrate of the blood–brain barrier is represented by microvascular endothelial cells. They are characterized by complexly arranged tight junctions and a very low rate of transcytotic vesicular transport. They express transport enzymes, carrier systems and brain endothelial cell-specific molecules of unknown function not expressed by any other endothelial cell population. These blood–brain barrier properties are not intrinsic to these cells but are inducible by the surrounding brain tissue. Type I astrocytes injected into the anterior eye chamber of the rat or onto the chick chorioallantoic membrane are able to induce a host-derived angiogenesis and some blood–brain barrier properties in endothelial cells of non-neural origin. Recently we have shown that this cellular interaction is due to the secretion of a soluble astrocyte derived factor(s). Astrocytes are also implicated in the maintenance, functional regulation and the repair of the blood–brain barrier. Complex interactions between other constituents of the microenvironment surrounding the endothelial cells, such as the basement membrane, pericytes, nerve endings, microglial cells and the extracellular fluid, take place and are required for the proper functioning of the blood–brain barrier, which in addition is regionally different as reflected by endothelial cell heterogeneity.

THE CONCEPT OF THE BLOOD–BRAIN BARRIER

The first generation of German pharmacologists, with Paul Ehrlich as prominent figure, performed at the end of the last century a number of experiments asking questions about the relation between the chemical structure of a given agent and its pharmacological action, as well as its tissue affinity and distribution. Although Paul Ehrlich himself described the observation that after intravenous application of some aniline dyes most of the animal tissues were stained with the exception of the central nervous system (CNS), he thought that this difference was due to different binding affinities (for review see Ehrlich 1902). The existence of a barrier at the level of cerebral vessels was first postulated by Biedl and Kraus (1898) and Lewandowsky (1900) based on the observation that the intravenous injection of cholic acids or sodium ferrocyanide had no pharmacological effects on the CNS, whereas neurological symptoms occurred after intraventricular application of the same substances. The existence of a blood–brain barrier was confirmed by the classic experiments of Goldmann (1909, 1913). By intravenous injections of trypan blue he reproduced the results of Ehrlich, showing that the whole animal turned blue with the exception of the CNS, but after subarachnoidal application he observed a selective staining of the CNS alone. Only later was it shown that the dyes used form complexes with serum proteins (Tschirgi 1950; Patterson et al 1992) and that in these early experiments it was essentially the barrier function for proteins that was demonstrated. Evidence for the concept of the blood–brain barrier was also provided by experiments using kinetic methods showing that the rapidity of penetration of a given substance into the brain depends on its lipophily (Davson and Matchett 1953). For a broader discussion of the older literature, see the reviews of Rapoport (1976) and Bradbury (1979).

MORPHOLOGICAL LOCALIZATION OF THE BLOOD–BRAIN BARRIER

Although on the basis of pharmacological and physiological evidence the concept of a blood–brain barrier was accepted, the question of its precise morphological localization remained controversial up to the end of the 1960s. Some early histological investigations reported that in experiments with systemic injections of dyes, where the brain tissue was not coloured, the endothelial cells of the brain were unstained too, whereas in stained tissues the endothelial cells were also stained (Spatz 1933; Bromann 1941). They concluded that the endothelial cells represented the site of the barrier. Early ultrastructural investigations, however, questioned the selective role of endothelial cells in the barrier function and the basement membrane and the surrounding astrocytic endfeet were claimed to be an essential part of the barrier (Dempsey and Wislocki 1955; VanBremen and Clemente 1955; Luse 1956; Gerschenfeld et al 1959).

In addition, another argument was advanced, since early ultrastructural studies could not confirm the presence of a significant extracellular space, postulated from physiological evidence (Davson and Spaziani 1959; Levin et al 1970), the whole concept of the blood–brain barrier was questioned (Gerschenfeld et al 1959). The very careful studies of VanHarrefeld et al (1965, 1966, 1967) set an end to this controversy. They showed that the lack of an extracellular space in the CNS in ultrastructural studies was due to the swelling of tissue components, especially of astrocytic endfeet before fixation. The old question of the morphological localization of the blood–brain barrier was, however, still not answered. The classical experiments to resolve this problem were done by Reese and Karnovsky (1967), Brightman and Reese (1969), Brightman and colleagues (1970a,b, 1971), Reese and colleagues (1971) and Bouldin and Krigman (1975). By using ultrastructural tracers like horseradish peroxidase, microperoxidase, colloidal or ionic lanthanum it could be shown that with intravenous as well as by intraventricular application the exchange of the tracers was stopped in both directions at the level of interendothelial tight junctions.

By freeze-fracture it could be shown (Nagy et al 1984; Shivers et al 1984) that the tight junctions between endothelial cells of CNS capillaries and venules are arranged in 6–8 parallel strands with complex net-like anastomoses all along the upper circumference of the endothelial cell. The complexity of the tight junctions is comparable to that observed in tight epithelia and restricts the passage of low-molecular-weight substances down to a diameter of 10–15 Å. This is also reflected by a very high transendothelial electrical resistance comparable to the values obtained in the skin or the urothel (Crone and Olesen 1982; Butt et al 1990).

Another characteristic of the blood–brain barrier endothelial cells is the very low rate of transcytotic vesicular transport as reflected by a very low content of intracytoplasmic vesicles compared, for example, to muscle capillaries (Coomber and Stewart 1985) and by the absence at the blood–brain barrier endothelial cells of the transcytosis-inducing sialoglycoprotein gp60 present in all other continuous capillaries (Schnitzer 1992).

With these two well-developed systems (tight junctions and low transcytotic transport) the CNS microvasculature controls the passive exchange of substances between blood and brain and represents the most effective barrier when compared to all other capillary beds. The content of proteins in the cerebrospinal fluid is only about 0.4% of that of serum and one has to take into account that there is even an additional contribution from periventricular regions without a blood–brain barrier.

MOLECULAR CHARACTERISTICS OF THE BLOOD–BRAIN BARRIER

The presence of a very efficient barrier for passive transport mechanisms made it necessary that the blood–brain barrier endothelial cells developed specific transport systems for important nutrients and for waste. In order to be able to selectively regulate the exchange between blood and brain or brain and blood of biologically important classes of substances, the blood–brain barrier endothelial cells express a number of enzymes, transporters and surface molecules of yet unknown function not significantly present in many other endothelial cell populations. These markers of the blood–brain barrier phenotype of endothelial cells include alkaline phosphatase (Goldstein and Harris 1981); nucleoside-diphosphatase (Vorbrodt et al 1983); abluminally localized $Na^+–K^+$-ATPase (Betz et al 1980); 5′-nucleotidase (Vorbrodt et al 1983); butylcholinesterase (Karcsu and Toth 1982); γ-glutamyl-transpeptidase (Albert et al 1966); aromatic L-amino acid decarboxylase (Stewart and Wiley 1981); guanylate cyclase (Karnushina et al 1980); aminopeptidase A (Juillerat-Jeanneret et al 1992); a transporter system for monocarboxy acids (Oldendorf 1973); a transporter for 16 neutral amino acids of the L-system (Pardridge 1984); a carrier for basic amino acids of the A-system (Betz and Goldstein 1978); a transporter for hexoses, including the glucose transporter (Dick et al 1984); and carriers for amines, nucleosides and purines (Pardridge 1984). Several antibodies to blood–brain barrier-specific epitopes have been developed (reviewed in Dermietzel and Krause 1991); these include the transferrin receptor (Jefferies et al 1984), a brain-specific epitope described by Auerbach et al (1985), the endothelial barrier antigen EBA (Rosenstein et al 1992), the P-glycoprotein (Cordon-Cardo et al 1989) and the membrane glycoprotein HT7/neurothelin (Seulberger et al 1990; Schlosshauer and Herzog 1990).

Probably the best-characterized blood–brain barrier marker is the glucose transporter (Dick et al 1984; Pardridge et al 1990; Farrell et al 1992; Dermietzel et al 1992). Six different types are expressed in a tissue- and cell-specific manner. The glucose transporter at the blood–brain barrier is the Glut-1 isoform. The molecular weight of this glycoprotein is 55 kDa. It has 12 transmembrane regions essential for its transport function and both the carboxy- and amino-terminal are located in the cytoplasm.

The most commonly used marker for the blood–brain barrier is γ-glutamyl-transpeptidase, an enzyme that mediates the transfer of the γ-glutamyl residue of glutathione to amino acids. It is a glycoprotein organized as heterodimer of a light (22 kDa) and a heavy (51 kDa) subunit. Both subunits originate from a common mRNA (Papandrikopoulou et al 1989). γ-Glutamyl-transpeptidase is expressed in many cell types, but in no other endothelial cells than those of the blood–brain barrier. It is also expressed in brain pericytes (Risau et al 1992).

A recently described blood–brain barrier-specific glycoprotein is the HT7 protein or neurothelin (Seulberger et al 1990; Schlosshauer 1990). It is a member of the immunoglobulin superfamily with two C2-type domains and a single transmembrane region corresponding to a leucine zipper structure. Its gene has been cloned in the chick and homologous proteins are known in the rat (OX-47) and in the mouse (gp42). Its exact function is not yet elucidated.

INTERACTIONS BETWEEN ASTROCYTES AND ENDOTHELIAL CELLS

Astrocytes surround with their endfeet 80–95% of the brain capillary circumference (Wolff 1963). They are implicated in the *de novo* induction, the maintenance and repair of the blood–brain barrier and in the angiogenesis of the microvasculature of the brain.

Transplantation experiments in which rat brain tissue was implanted into the anterior eye chamber of syngeneic adult rats (Svendgaard et al 1975), or in which allograft or xenograft fetal rat brain cell suspensions were implanted into adult rat brain (Geist et al 1991), or in which embryonic chick or quail brain was implanted into the coelomic cavity in the chick–quail chimera model (Stewart and Wiley 1981), or in which embryonic mouse brain was transplanted onto the chick chorioallantoic membrane (Risau et al 1986; Schlosshauer and Herzog 1990) suggested that the specific properties of the endothelial cells of the blood–brain barrier phenotype were not intrinsic to these cells but induced by the surrounding brain tissue. Some blood–brain barrier markers, such as non-permeability to Evans blue bound to albumin and expression of tight junctions and of alkaline phosphatase have been shown to be inducible by astrocytes in endothelial cells of non-neural origin (Janzer and Raff 1987; Shivers et al 1988; Tio et al 1990). We have recently shown that a soluble factor derived from fetal rat astrocytes is able to induce the expression of the blood–brain barrier-specific HT7 protein/neurothelin in chick chorioallantoic vessels (Janzer et al 1993; Lobrinus et al 1992).

Astrocytes are also implicated in the maintenance of the differentiated blood–brain barrier phenotype of brain endothelial cells. Co-cultures of brain-derived endothelial cells with astrocytes or exposure to astrocyte-conditioned medium increase or re-induce blood–brain barrier properties lost after a few passages *in vitro*, such as induction of tight junctions or increase in transendothelial electrical resistance (Tao-Cheng et al 1987; Dehouk et al 1990; Raub et al 1992), of γ-glutamyl-transpeptidase expression (DeBault and Cancilla 1979; Maxwell et al 1987; Dehouk et al 1990), of glucose uptake (Maxwell et al 1989), of Na^+-K^+-ATPase and non-specific alkaline phosphatase expression (Beck et al 1986) and of leucine uptake (Juillerat-Jeanneret et al 1993).

Microvessel formation *in vitro* and probably during normal development is also influenced by astrocytes (Wolff et al 1992b).

There is increasing evidence that not only astrocytes but several other components of the microenvironment of the endothelial cells at the blood–brain barrier influence and regulate the function of the blood–brain barrier. These include the basement membrane (Shivers et al 1988; Augustin-Voss et al 1991; Juillerat-Jeanneret et al 1993), the pericytes (Risau et al 1992), the neuronal population (Tontsch and Bauer 1991) as well as microglial cells and the extracellular fluid. It is also noteworthy that the regulation of permeability and of metabolic markers such as glucose transport is dissociated, at least under pathological conditions (Guerin et al 1990). In addition there is increasing evidence that there is heterogeneity at the level of the blood–brain barrier endothelial cells in respect to regional and segmental distribution (Lasbennes et al 1985; Owman and Hardebo 1988; Bauer et al 1990; Wolff et al 1992a; Juillerat-Jeanneret et al 1993).

CONCLUSIONS

The cellular basis of the blood–brain barrier is well established and is localized at the level of the brain microvascular endothelial cells. The proper barrier function is mediated by tight junctions and the virtual absence of transcytosis. The barrier function is induced, maintained, regulated and repaired by astrocytes. In addition, the endothelial cells of the blood–brain barrier have a number of specific metabolic and molecular characteristics. These are regionally different and are under separate control of astrocyte-derived factors. Recent evidence for endothelial cell and astrocyte heterogeneity at the blood–brain barrier and for the important and not fully understood role of other than astrocytic components of the microenvironment of the blood–brain barrier, such as the basement membrane, pericytes, neuronal cells, microglial cells and the extracellular fluid composition, for the proper function and regulation of the blood–brain barrier indicates that the blood–brain barrier, although confirmed as a valid concept, is more complex than thought 10 years ago.

REFERENCES

Albert Z, Orlowski M, Ruzucidlo P, Orlowski J (1966) Studies on gamma-glutamyl-transpeptidase activity and its histochemical localisation in the central nervous system of men and different animal species. *Acta Histochem (Suppl.)* **25**: 312–320.

Auerbach R, Alby L, Morrissey LW, Tu M, Joseph J (1985) Expression of organ-specific antigens on capillary endothelial cells. *Microvasc Res* **29**: 401–411.

Augustin-Voss HG, Johnson RC, Pauli BU (1991) Modulation of endothelial cell surface glycoconjugate expression by organ derived biomatrices. *Exp Cell Res* **192**: 346–351.

Bauer HC, Tontsch U, Amberger A, Bauer H (1990) Gamma-glutamyl-transpeptidase and Na^+,K^+-ATPase activities in different subpopulations of cloned cerebral endothelial cells: responses to glial stimulation. *Biochem Biophys Res Commun* **168**: 358–363.

Beck DW, Roberts RL, Olson JJ (1986) Glial cells influence membrane-associated enzyme activity at the blood–brain barrier. *Brain Res* **43**: 131–137.

Betz AL, Goldstein GW (1978) Polarity of the blood–brain barrier: neutral amino acid transport into isolated brain capillaries. *Science* **202**: 225–226.

Betz AL, Firth JA, Goldstein GW (1980) Polarity of the blood–brain barrier: distribution of enzymes between the luminal and antiluminal membranes of brain capillary endothelial cells. *Brain Res* **192**: 17–28.

Biedl A, Kraus R (1898) Über eine bisher unbekannte toxische Wirkung der Gallensäuren auf das Zentralnervensystem. *Zbl Inn Med* **19**: 1185–1200.

Bouldin TW, Krigman MR (1975) Differential permeability of cerebral capillary and choroid plexus to lanthanum ion. *Brain Res* **99**: 444–448.

Bradbury MWB (1979) *The Concept of a Blood–Brain Barrier*. Chichester: Wiley.

Brightman MW, Reese TS (1969) Junctions between intimately apposed cell membranes in the vertebrate brain. *J Cell Biol* **40**: 648–677.

Brightman MW, Klatzo I, Olsson Y, Reese TS (1970a) The blood–brain barrier to proteins under normal and pathological conditions. *J Neurol Sci* **10**: 215–239.

Brightman MW, Reese TS, Feder N (1970b) Assessment with the electron microscope of the permeability to peroxidase of cerebral endothelium in mice and sharks. In Crone C, Lassen NA, eds. *Capillary Permeability*. New York: Academic Press, 463–476.

Brightman MW, Reese TS, Olsson Y, Klatzo I (1971) Morphologic aspects of the blood–brain barrier to peroxidase in elasmobranchs. In Zimmermann HM, ed. *Progress in Neuropathology*, Vol 1. New York: Grune & Stratton, 146–161.

Broman T (1941) The possibilities of the passage of substances from the blood to the central nervous sysstem. *Acta Psychiatr* **16**: 1–25.

Butt AM, Jones HC, Abbott NJ (1990) Electrical resistance across the blood–brain barrier in anesthetized rats: a developmental study. *J Physiol* **429**: 47–62.

Coomber BL, Stewart PA (1985) Morphometric analysis of CNS microvasculature endothelium. *Microvasc Res* **30**: 99–115.

Cordon-Cardo C, O'Brien JP, Casals D et al (1989) Multidrug-resistance gene (P-glycoprotein) is expressed by endothelial cells at the blood–brain barrier. *Proc Natl Acad Sci USA* **86**: 695–998.

Crone C, Oleson SP (1982) Electrical resistance of brain microvasculature endothelium. *Brain Res* **241**: 49–55.

Davson H, Matchett PA (1953) The kinetics of penetration of the blood–aqueous barrier. *J Physiol* **122**: 11–32.

Davson H, Spaziani E (1959) The blood–brain barrier. *J Physiol* **149**: 135–143.

DeBault LE, Cancilla PA (1979) Gamma-glutamyl-transpeptidase in isolated brain endothelial cells: induction by glial cells in vitro. *Science* **207**: 653–655.

Dehouk MP, Meresse S, Delorme P, Fruchart JC, Cecchelli R (1990) An easier, reproducible, and mass production method to study the blood–brain barrier in vitro. *J Neurochem* **54**: 1789–1801.

Dempsey EW, Wislocki GB (1955) An electron microscopic study of the blood–brain barrier in the rat, employing silver nitrate as a vital stain. *J Biophys Biochem Cytol* **1**: 245–256.

Dermietzel R, Krause D (1991) Molecular anatomy of the blood–brain barrier as defined by immunocytochemistry. *Int Rev Physiol* **127**: 57–109.

Dermietzel R, Krause D, Kremer M, Wang G, Stevenson B (1992) Pattern of glucose transporter (glut-1) in embryonic brains is related to maturation of blood–brain barrier tightness. *Dev Dyn* **193**: 152–163.

Dick APK, Harik SI, Klip A, Walker DM (1984) Identification and characterization of the glucose transporter of the blood–brain barrier by cytochalasin B binding and immunological reactivity. *Proc Natl Acad Sci USA* **81**: 7233–7237.

Ehrlich P (1902) *Über die Beziehung von chemischer Constitution, Vertheilung und pharmakologischer Wirkung*. Berlin: Hirschwald.

Farrell CL, Yang J, Pardridge WM (1992) Glut-1 transporter is present within apical and basolateral membranes of brain epithelial interfaces and in microvascular endothelia with and without tight junctions. *J Histochem Cytochem* **40**: 193–199.

Geist MJ, Maris DO, Grady MS (1991) Blood–brain barrier permeability is not altered by allograft or xenograft fetal neural cell suspension grafts. *Exp Neurol* **111**: 166–174.

Gerschenfeld HM, Wald F, Zadunaisky JA, DeRobertis EDP (1959) Function of astroglia in the water–ion metabolism of the central nervous system. An electron microscope study. *Neurology* **9**: 412–425.

Goldmann EE (1909) Die äussere und innere Sekretion des gesunden und kranken Organismus im Lichte der 'vitalen Färbung'. *Beitr Z Klin Chir* **64**: 192–265.

Goldmann EE (1913) *Vitalfärbungen am Zentralnervensystem. Beitrag zur Physiologie des Plexus choroideus und der Hirnhäute.* Berlin: Hirschfeld.

Goldstein TR, Harris H (1981) Mammalian brain alkaline phosphatase: expression of liver/bone/kidney locus. *J Neurochem* **36**: 53–57.

Guerin C, Laterra J, Hruban RH, Brem H, Drewes LR (1990) The glucose transporter and blood–brain barrier of human brain tumors. *Ann Neurol* **28**: 758–765.

Janzer RC, Raff MC (1987) Astrocytes induce blood–brain barrier properties in endothelial cells. *Nature* **325**: 253–257.

Janzer RC, Lobrinus A, Juillerat-Jeanneret L, Darekar P (1993) A soluble astrocytic factor induces the expression of HT7 and neurothelin in endothelial cells of the chick chorioallantoic vessels. In Drewes L, Betz LA, eds. *Frontiers in Cerebral Vascular Biology: Transport and its Regulation.* New York: Plenum Press, in press.

Jefferies WA, Brandon MR, Hunt SV, Williams AF, Gatter KC, Mason DY (1984) Transferrin receptor on endothelium of brain capillaries. *Nature* **312**: 162–163.

Juillerat-Jeanneret L, Aguzzi A, Wiestler OD, Darekar P, Janzer RC (1992) Dexamethasone selectively regulates the activity of enzymatic markers of cerebral endothelial cell lines. *In Vitro Cell Develop Biol* **28A**: 537–543.

Juillerat-Jeanneret L, Darekar P, Janzer RC (1993) Heterogeneity of microvascular endothelial cells of the brain: a comparison of the effects of extracellular matrix and soluble astrocytic factors. *Endothelium: J Endothel Cell Res* in press.

Karcsu S, Toth L (1982) Die Veränderungen der Butyryl-Cholinesterase-Akitivität der fenestrierten Kapillaren in der Area postrema während der postnatalen Entwicklung. *Acta Histochem* **71**: 83–94.

Karnushina IL, Toth I, Dux E, Joo F (1980) Presence of guanylate cyclase in brain capillaries: histochemical and biochemical evidence. *Brain Res* **189**: 588–596.

Lasbennes F, Sercombe R, Verrechia C, Seylaz J (1985) Vascular monoamine oxidase activity in the rat brain: variation with the substrate and the vascular segment. *Life Sci* **36**: 2263–2268.

Levin VA, Fenstermacher JD, Patlak CS (1970) Sucrose and insulin space measurements of cerebral cortex in four mammalian species. *Am J Physiol* **219**: 1528–1533.

Lewandowsky M (1990) Zur Lehre der Zerebrospinalflüssigkeit. *Z Klin Med* **40**: 480–494.

Lobrinus A, Juillerat-Jeanneret L, Darekar P, Schlosshauer B, Janzer RC (1992) Induction of the blood–brain barrier specific HT7 and neurothelin epitopes in endothelial cells of the chick chorioallantoic vessels by a soluble astrocyte derived factor. *Dev Brain Res* **70**: 207–211.

Luse SA (1956) Electron microscopic observations of the central nervous system. *J Biophys Biochem Cytol* **2**: 531–542.

Maxwell K, Berliner JA, Cancilla PA (1987) Induction of gamma-glutamyl-transpeptidase in cultured cerebral endothelial cells by a product released by astrocytes. *Brain Res* **410**: 309–314.

Maxwell K, Berliner JA, Cancilla PA (1989) Stimulation of glucose analogue uptake by cerebral microvessel endothelial cells by a product released by astrocytes. *J Neuropathol Exp Neurol* **48**: 69–80.

Nagy Z, Peters H, Hüttner I (1984) Fracture faces of cell junctions in cerebral endothelium during normal and hyperosmotic conditions. *Lab Invest* **50**: 313–322.

Oldendorf WH (1973) Carrier-mediated blood–brain barrier transport of short-chain monocarboxylic organic acids. *Am J Physiol* **224**: 1450–1453.

Owman C, Hardebo JE (1988) Functional heterogeneity of cerebrovascular endothelium. *Brain Behav Evol* **32**: 65–75.

Papandrikopoulou A, Frey A, Gassen HG (1989) Cloning and expression of gamma-glutamyl-transpeptidase from isolated porcine brain capillaries. *Eur J Biochem* **183**: 693–698.

Pardridge WM (1984) Transport of nutrients and hormones through the blood–brain barrier. *Fedn Proc* **43**: 201–204.

Pardridge WM, Boado RJ, Farrell CR (1990) Brain-type glucose transporter (Glut-1) is selectively localized to the blood–brain barrier. Studies with quantitative western blotting and in situ hybridization. *J Biol Chem* **265**: 18035–18040.

Patterson CE, Rhoades RA, Garcia JG (1992) Evans blue dye as a marker of albumin clearance in cultured endothelial monolayer and isolated lung. *J Appl Physiol* **72**: 865–873.

Raub TJ, Kuentzel SL, Sawada GA (1992) Permeability of bovine brain microvessel endothelial cells in vitro: barrier tightening by a factor released from astroglioma cells. *Exp Cell Res* **199**: 330–340.

Rapoport SI (1976) *Blood–brain Barrier in Physiology and Medicine*. New York: Raven Press.

Reese TS, Karnovsky MJ (1967) Fine structural localisation of a blood–brain barrier to exogenous peroxidase. *J Cell Biol* **34**: 207–217.

Reese TS, Feder N, Brightman MWQ (1971) Electron microscopic study of the blood–brain barrier and cerebrospinal fluid barriers with microperoxidase. *J Neuropathol Exp Neurol* **30**: 137–148.

Risau W, Hallmann R, Albrecht U, Henke-Fahle S (1986) Brain induces the expression of an early cell surface marker for blood–brain barrier specific endothelium. *EMBO J* **5**: 3179–3183.

Risau W, Dingler A, Albrecht U, Dehouk MP, Cecchelli R (1992) Blood–brain barrier pericytes are the main source of gamma-glutamyl-transpeptidase activity in brain capillaries. *J Neurochem* **58**: 667–672.

Rosenstein JM, Krum JM, Sternberger LA, Pulley MT (1992) Immunocytochemical expression of the endothelial barrier antigen (EBA) during brain angiogenesis. *Dev Brain Res* **66**: 47–54.

Schlosshauer B (1991) Neurothelin: molecular characteristics and developmental regulation in the chick CNS. *Development* **113**: 129–140.

Schlosshauer B, Herzog KH (1990) Neurothelin: an inducible cell surface glycoprotein of blood–brain barrier specific endothelial cells and distinct neurons. *J Cell Biol* **110**: 1261–1274.

Schnitzer JE (1992) gp60 is an albumin-binding glycoprotein expressed by continuous endothelium involved in albumin transcytosis. *Am J Physiol* **262**: P54.

Seulberger H, Lottspeich F, Risau W (1990) The inducible blood–brain barrier specific molecule HT7 is a novel immunoglobulin-like cell surface glycoprotein. *EMBO J* **9**: 2151–2158.

Shivers RR, Betz AL, Goldstein GW (1984) Isolated rat brain capillaries possess intact, structurally complex, interendothelial tight junctions: freeze-fracture verification of tight junction integrity. *Brain Res* **324**: 313–322.

Shivers RR, Arthur FE, Bowman PD (1988). Induction of gap junctions and brain-endothelium-like tight junctions in cultured bovine endothelial cells: local control of cell specialization. *J Submicrosc Cytol Pathol* **20**: 1–14.

Spatz H (1933) Die Bedeutung der vitalen Färbung für die Lehre vom Stoffaustausch zwischen dem Zentralnervensystem und dem übrigen Körper. *Arch Psychiatr Nervenkr* **101**: 267–358.

Stewart PA, Wiley MJ (1981) Developing nervous tissue induces formation of blood–brain characteristics in invading endothelial cells: a study using quail–chick transplantation chimeras. *Dev Biol* **84**: 183–192.

Svendgaard NA, Björklund A, Hardebo A, Stenevi U (1975) Axonal degeneration associated with a defective blood–brain barrier in cerebral implants. *Nature* **255**: 334–337.

Tao-Cheng JH, Nagy Z, Brightman MW (1987) Tight junctions of brain endothelium in vitro are enhanced by astroglia. *J Neurosci* **7**: 3293–3299.

Tio S, Deenen M, Marani E (1990) Astrocyte-mediated induction of alkaline phosphatase activity in human umbilical cord vein endothelium: an in vitro model. *Eur J Morphol* **28**: 289–300.

Tontsch U, Bauer HC (1991) Glial cells and neurons induce blood–brain barrier related enzymes in cultured cerebral endothelial cells. *Brain Res* **539**: 247–253.

Tschirgi RD (1950) Protein complexes and the impermeability of the blood–brain barrier. *Am J Physiol* **163**: 756–758.

VanBremen VL, Clemente CD (1955) Silver deposition in the central nervous system and the hematoencephalic barrier studied with the electron microscope. *J Biophys Biochem Cytol* **1**: 161–198.

VanHarrefeld A, Malhotra SK (1967) Extracellular space in the cerebral cortex of the mouse. *J Anat* **101**: 197–207.

VanHarrefeld A, Crowell J, Malhotra SK (1965) A study of extracellular space in central nervous tissue by freeze substitution. *J Cell Biol* **25**: 117–137.

VanHarrefled A, Collewijn H, Malhotra SK (1966) Water, electrolyte and extracellular space in hydrated and dehydrated brains. *Am J Physiol* **210**: 251–256.

Vorbrodt AW, Lossinsky AS, Wisniewski HM (1983) Enzyme cytochemistry of blood–brain barrier disturbances. In Hossmann KA, Klatzo I, eds. *Cerebrovascular Transport Mechanisms.* Acta Neuropathol. (Suppl VIII). Berlin: Springer, 43–59.

Wolff J (1963) Beiträge zur Ultrastruktur der Capillaren der normalen Grosshirnrinde. *Z Zellforsch* **60**: 409–431.

Wolff JE, Belloni-Olivi L, Bressler JP, Goldstein GW (1992a) Gamma-glutamyl-transpeptidase activity in brain microvessels exhibits regional heterogeneity. *J Neurochem* **58**: 909–915.

Wolff JE, Laterra J, Goldstein GW (1992b) Steroid inhibition of neural microvessel morphogenesis in vitro: receptor mediation and astroglial dependence. *J Neurochem* **58**: 1023–1032.

J. Inher. Metab. Dis. 16 (1993) 648–669
© SSIEM and Kluwer Academic Publishers. Printed in the Netherlands

Physiology and Pathophysiology of Organic Acids in Cerebrospinal Fluid

G. F. Hoffmann[1,2]*, W. Meier-Augenstein[1], S. Stöckler[3], R. Surtees[4], D. Rating[1] and W. L. Nyhan[2]

Departments of Pediatrics, [1]University of Heidelberg, Germany, [2]University of California, San Diego, USA, and [3]University of Graz, Austria; [4]Institute of Child Health, London, UK

Summary: Concentrations of organic acids in cerebrospinal fluid (CSF) appear to be directly dependent upon their rate of production in the brain. There is evidence that the net release of short-chain monocarboxylic acids from the brain is a major route for removing these products of cerebral metabolism. Concentrations of organic acids in blood and CSF are largely independent of each other. Quantitative reference values for the concentrations of organic acids in CSF and plasma as well as ratios of individual organic acids between CSF and plasma were determined in 35 pairs of samples from paediatric patients. Over 25 organic acids were quantifiable in all or in the majority of CSF and/or plasma specimens (limit of detection 1 μmol/L). There were substantial differences in the CSF/plasma ratios between subgroups of organic acids. Metabolites related to fatty-acid oxidation were present in CSF in substantially less amounts than in plasma. Organic acids related to carbohydrate and energy metabolism and to amino acid degradation were present in CSF in the same amounts as or slightly smaller amounts than in plasma. Finally, some organic acids were found in substantially higher amounts in CSF than in plasma, e.g. glycolate, glycerate, 2,4-dihydroxybutyrate, citrate and isocitrate.

Studies of organic acids in CSF and plasma samples are presented from patients with 'cerebral' lactic acidosis, disorders of propionate and methylmalonate metabolism, glutaryl-CoA dehydrogenase deficiency and L-2-hydroxyglutaric acidura. It became apparent that derangements of organic acids in the CSF may occur independently of the systemic metabolism. Quantitative organic acid analysis in CSF will yield new information on the pathophysiology in the central nervous system (CNS) of these disorders and may prove necessary for successful monitoring of treatment of organoacidopathies, which present mainly with neurological disease. For example, in glutaryl-CoA dehydrogenase deficiency the urinary excretion of glutarate appears to be an inadequate parameter for monitoring the effect of dietary therapy, without plasma and

---

*Correspondence: G. F. Hoffmann, Universitäts-Kinderklinik, Heidelberg, Im Neuenheimer Feld 150, D-6900 Heidelberg, Germany

CSF determinations. In L-2-hydroxyglutaric aciduria the elevation of L-2-hydroxyglutarate was found to be greater in CSF than in plasma. In addition, some other organic acids, glycolate, glycerate, 2,4-dihydroxybutyrate, citrate and isocitrate, were also elevated in the CSF of the patients out of proportion to normal levels in plasma and urine. High concentrations of an unknown compound, which was tentatively identified as 2,4-dihydroxyglutarate, were found in the CSF of patients with L-2-hydroxyglutaric aciduria. Quantitative determination of organic acids in CSF and plasma should aid the monitoring of treatment of patients with organic acid disorders, allow investigations of metabolites in known disorders, and detect neurometabolic diseases in which the diagnostic metabolites accumulate preferentially in CSF.

The inborn errors of amino and organic acid metabolism have been rich resources for the study of the effects of altered biochemistry on the CNS. Much, however, remains to be learned. It has become increasingly apparent that disorders of organic acid metabolism are responsible for a wide spectrum of neurological disease (Ozand and Gascon 1991). Neurological manifestations are well-established elements of the phenotypes of disorders such as methylmalonic aciduria (Andreula et al 1991), propionic aciduria (Sethi et al 1989) or glutaryl-CoA dehydrogenase deficiency (Amir et al 1989; Haworth et al 1991; Hoffmann et al 1991). In addition, patients have been described with primary CNS 'organic acidurias' (Roe et al 1988; Barth et al 1992) as well as 'cerebral' lactic acidosis (Brown et al 1988; Benninger et al 1990). Insight into these and other such disorders will require accurate, sensitive and specific quantitative determinations of organic acids in CSF.

In the first part of this paper some general principles of the metabolism of organic acids in the CNS and especially of their transport across the blood–brain–CSF barrier are summarized. The second part deals with analytical methods and control values. Finally, we report on literature data and personal experience with analyses of organic acids in CSF in organoacidopathies and neurometabolic disorders.

NEUROCHEMISTRY OF ORGANIC ACIDS AND THEIR TRANSPORT ACROSS THE BLOOD–BRAIN–CSF BARRIER

In recent years, some specific organic acids in CSF and CNS have been the subject of a number of clinical and experimental studies (for overviews see Rapoport 1976; Fishman 1980). Lactate has been studied most extensively. This is because of the relative ease of its accurate determination, its diagnostic and prognostic value in diseases such as cerebral infarction, meningitis and other meningeal disorders associated with depressed CSF glucose levels, and its potential role in the regulation of important physiological functions, such as pulmonary ventilation, cerebral blood flow and metabolism (Weyne and Leusen 1975; Rapoport 1976; Knudsen et al 1991). However, it is difficult to extrapolate the findings of these studies to the physiology of organic acids as a group. Nevertheless, some principles, especially about the transport of organic acids across the blood–brain–CSF barrier, appear to be established.

J. Inher. Metab. Dis. 16 (1993)

There is substantial evidence in humans as well as in experimental animals that the concentrations of organic acids in blood and CSF are largely independent of each other. Most, if not all, organic acids in CSF appear to be directly dependent upon their rate of production in the brain. Organic acids diffuse freely from brain to CSF (Fenstermacher and Patlak 1975), and the concentrations in CSF are closely linked to the respective concentrations in brain tissue (Weyne and Leusen 1975; Fishman 1980). There is evidence that the net release of short-chain monocarboxylic acids from brain is a major route for removing these products of cerebral metabolism, because the concentrations of organic acids in CSF are slightly lower than those in the brain following diffusion equilibrium. Organic acids are removed from CSF by active transport at the chorioid plexus and by bulk flow. Organic acids are cleared from CSF at a minimum rate of 25% per hour by bulk removal alone, because drainage occurs through membranes of the arachnoid villi that do not discriminate among molecules of different sizes and the turnover rate of CSF by bulk flow is approximately 25% per hour (Cutler et al 1968).

Short-chain monocarboxylic organic acids cross cerebral capillaries by a common carrier-mediated facilitated mechanism (Oldendorf 1973). It seems likely that no fundamental differences exist in the mechanism of transport at the blood–brain and blood–CSF interfaces. The system transfers L-lactate, pyruvate, acetate, propionate, butyrate (Oldendorf 1973), and probably D-3-hydroxybutyrate and acetoacetate (Cremer et al 1976). No measurable transport was found of the di- or tricarboxylic acids or of p-aminohippurate. The organic acid carrier system can be completely inhibited, is saturable, stereospecific, and is independent of amino acid or glucose transport. The carrier transport system is probably bi-directional; as has been described for amino acids or sugars. In most situations, the outward brain-to-blood transport of lactate is accompanied by influx of 3-hydroxybutyrate and acetoacetate (Cremer et al 1976). The transport capacity of the carrier is relatively low, the estimated K_m value for lactate is $\approx 2\,\text{mmol/L}$ and for pyruvate is $\geq 0.5\,\text{mmol/L}$ (Oldendorf 1973; Pardridge et al 1975). Saturation of influx at moderately elevated plasma concentrations can be viewed as a mechanism to protect the brain in cases of systemic lactic acidosis. However, when lactate and pyruvate accumulate in the brain under conditions of hypoxia or in congenital cerebral lactic acidosis, the low-capacity transport system cannot clear the excess metabolites into the blood and intracerebral accumulation of short-chain monocarboxylic acids is likely to occur (Weyne and Leusen 1975). Saturation will be approached at concentrations of $\geq 4\,\text{mmol/L}$ in CSF in the case of L-lactate, which is often reached in congenital lactic acidosis.

Monocarboxylic acids whose chain lengths exceed that of butyric acid, and other organic acids, enter the brain by passive diffusion on the basis of increasing lipid solubility. This uptake is virtually total at lengths greater than that of hexanoate (Oldendorf 1973; Rapoport 1976), shows no cross-competitive effect with short-chain monocarboxylic acids, and is not saturable. Equilibrium blood/CSF concentration ratios of organic acids are further influenced by the pH difference between CSF and blood. In normal subjects the arterial pH is about 7.40 and the CSF about 7.32. This small pH difference across the barrier partly excludes weak acids from the CSF (compartment with lower pH) and favours entry into the blood (compartment with higher pH) (Rall et al 1959). This effect is significant for organic acids with a high pK_a (≤ 5). During severe systemic acidosis, as seen in the organoacidopathies, the pH distribution is reversed, and organic acids may actually accumulate in the CSF and brain (compartment with higher pH). In general, the removal of organic acids by the CSF route becomes more important in pathological states, when increased amounts of organic acids are produced in the brain. For example, the accumulation of plasma organic acids in uraemia may inhibit the transport systems of the choroid plexuses, the arachnoid villi, and the capillary endothelium, resulting in disturbances of the excretory function of the CSF. This has been implicated in the pathogenesis of uraemic encephalopathy (Fishman 1980).

REFERENCE VALUES FOR ORGANIC ACIDS IN CSF

In contrast to a number of quantitative studies on amino acids in CSF (e.g. Gerrits et al 1989), which have been facilitated by the availability of automated amino acid analysers for the separation and quantitation of the free amino acids in protein-free filtrates since the 1950s, reference ranges do not exist for ninhydrin-negative organic acids in CSF by 'general' organic acid analysis (Chalmers and Lawson 1982; Sweetman 1991). The data currently available come in case reports or reports concerning specific inborn errors of metabolism, where the control populations chosen are frequently inappropriately small in size and poorly defined in terms of 'normality'. On the other hand, a number of clinical and experimental studies, but focusing on only one or a few specific organic acids, report control ranges of some organic acids in CSF and CNS. For some metabolites, stable isotope dilution selective-ion monitoring techniques have become the methods of choice for accurate quantification of single target compounds. Besides lactate and pyruvate, organic acids studied in more detail in CSF (with normal ranges, developmental variations and disturbances in neurological, psychiatric and metabolic disorders) include the metabolites of neurotransmitters such as homovanillate derived from dopamine, 5-hydroxyindole-acetate from 5-hydroxytryptamine, phenylacetate from 2-phenylethylamine and 2-oxoglutaramate from glutamine. The fatty-acid composition of CSF has also been studied by a number of investigators. No clinically significant differences between controls and patients with neurological and psychiatric disorders have been apparent (Müller and Vahar-Martiar 1975; Tichy and Skorkovska 1979).

It was not until small, easy-to-use bench-top mass spectrometers became available that organic acids in CSF could be studied more systematically. Recently, organic acid profiles in CSF have been investigated in the framework of a protocol designed to establish the frequency of inborn errors of metabolism in SIDS (Divry et al 1990; Coude et al 1991; Coude and Kamoun 1992). In these studies only four organic acids were constantly found in control CSFs, namely 2-hydroxybutyrate, 3-hydroxybutyrate, palmitate and stearate. An additional eight acids were inconstantly present. After death, the concentrations of most acids increased and eight additional metabolites, not observed in CSF from live infants, became detectable. For three of them, malate, lactyllactate and uracil, concentrations increased with the delay of sampling after death (Coude and Kamoun 1992).

Because reference ranges for organic acids in CSF by 'general' organic acid analysis have not been available, we employed an improved and more sensitive method for quantitative organic acid analysis (Hoffmann et al 1989; Sweetman 1991) to determine the concentrations of organic acids in CSF and plasma and their ratios in a paediatric population. Increased precision and sensitivity were possible through a combination of four features. First, known amounts of standards, more than 170 at present, were carried through the entire method to obtain multi-point standard curves. The most critical step was a quantitative isolation of as wide a range of organic acids as possible, which was achieved by batchwise liquid partition chromatography. Automated identification and quantification of compounds became feasible with the availability of benchtop GCMS and appropriate software. Finally, the sensitivity and accuracy of the quantification of oxo acids was greatly improved by the formation of O-(2,3,4,5,6-pentafluorobenzyl)oximes (Hoffmann and Sweetman 1987). These derivatives uniformly exhibit the pentafluorotropylium ion $[C_6F_5CH_2]^+$ at $m/z = 181$, accounting for 10–55% of the total ion current and making it ideally suited for identification and quantification (Hoffmann and Sweetman 1991; Hoffmann et al 1989).

Thirty-five paired control specimens of CSF and plasma were chosen retrospectively from more than 1000 pairs of specimens of CSF and plasma, which were obtained simultaneously from paediatric patients undergoing a diagnostic lumbar puncture during the investigation of suspected meningitis, Lyme disease or other neurological disorders. The reference group consisted of 15 girls and 20 boys, aged 0.2 to 16 years. In these patients there was no evidence of malignant disease or inborn error of metabolism. Immunological studies gave no evidence of Lyme disease. The CSF erythrocyte content was $< 100/\mu l$, the number of white blood cells was $< 10/\mu l$, the protein content was $< 0.5\,g/L$, and glucose concentrations were normal. Table 1 lists the reference values for 40 organic acids in CSF and plasma. An organic acid chromatogram of a controL CSF is shown in Figure 1.

Over twenty-five compounds were present and quantifiable in the majority of CSF and/or plasma specimens (Table 1). Ratios of organic acids between CSF and plasma are listed in Table 2. As the limit for reliable quantification for compounds by this method is around 1–3 nmol and most metabolites of interest in CSF and plasma occur in the low micromolar range, the ideal sample volume is 1 ml. The accuracy of quantification in the low micromolar range is around 20%. It is, however, possible to analyse samples down to 0.5 ml of CSF or plasma and still obtain sufficient sensitivity for quantitative evaluation. In contrast to concentrations of metabolites of neurotransmitters such as homovanillic acid, 5-hydroxyindoleacetic acid and 3-O-methyldopa, which were also determined in the same set of samples and were significantly higher in infants < 1 year, no obvious age- or sex-differences could be detected for any organic acid. This was unexpected, as most amino acids are found to be higher in younger children (Gerrits et al 1989). Furthermore, there is a clear age-dependency in the urinary excretion of organic acids, even beyond the first year of life. In interpreting the results of our analysis of organic acids in CSF, one has to bear in mind that the numbers of samples in different age groups were limited; and it may be appropriate to re-investigate single acids to detect more subtle age-dependencies.

Table 1   Concentrations of organic acids in control CSF and plasma samples

	Cerebrospinal fluid			Plasma		
	Mean (% Occ.)	Min.	Max.	Mean (% Occ.)	Min.	Max.
Total organic acids (meq/L)	10.4 (100%)	6	15	12.2 (100%)	5	19
Compound (μmol/L)						
Lactate	850 (100%)	450	2100	1700 (100%)	700	3300
2-Hydroxyisobutyrate	nd (0%)	nd	nd	7 (13%)	nd	9
Hexanoate	trace (32%)	nd	1.5	17 (75%)	nd	105
Glycolate	54 (100%)	5	250	27 (100%)	9	42
2-Hydroxybutyrate	35 (100%)	11	86	54 (100%)	8	80
3-Hydroxypropionate	4.4 (50%)	nd	9.5	2.3 (40%)	nd	4
3-Hydroxybutyrate	48 (100%)	trace	280	180 (100%)	22	700
3-Hydroxyisobutyrate	18 (93%)	trace	38	20 (100%)	4	48
2-Hydroxyisovalerate	6.8 (75%)	nd	18	7.7 (80%)	nd	19
Octanoate	2.5 (3%)	nd	2.5	8 (32%)	5	19
Succinate	3 (25%)	nd	5	9 (73%)	nd	32
Glycerate	34 (100%)	trace	95	10 (93%)	nd	24
Fumarate	trace (5%)	nd	trace	1.5 (78%)	nd	4
Glutarate	trace (5%)	nd	trace	0.8 (17%)	nd	1.8
2,4-Dihydroxybutyrate	84 (44%)	nd	260	2 (40%)	nd	7
3,4-Dihydroxybutyrate	15 (28%)	nd	73	18 (50%)	nd	54
Decanoate	nd (0%)	nd	nd	11 (10%)	5	17
Malate	3 (13%)	nd	4.5	12 (78%)	nd	21
Pyruvate	71 (95%)	trace	102	92 (100%)	27	160
Pyroglutamate	41 (100%)	10	96	51 (100%)	13	161
2-Oxoisovalerate	8.2 (63%)	nd	15	14 (72%)	nd	28
Erythronate	5 (30%)	nd	21	2 (13%)	nd	5
2-Hydroxyglutarate	1 (15%)	nd	3	1.5 (6%)	nd	1.5
Acetoacetate	6 (84%)	nd	32	21 (95%)	nd	86
2-Oxo-3-methyl-n-valerate	2 (77%)	nd	8	18 (100%)	8	31
2-Oxoisocaproate	5 (93%)	nd	9.4	28 (93%)	nd	58
Laurate	2.8 (64%)	nd	6.3	12 (100%)	2	37
Suberate	1.7 (3%)	nd	1.7	3.6 (8%)	nd	10
Aconitate	2 (5%)	nd	4	nd (0%)	nd	nd
Azelate	17 (17%)	nd	35	27 (44%)	nd	58
Isocitrate	10 (100%)	1	22	6 (90%)	nd	10
Citrate	350 (100%)	90	590	190 (100%)	30	400
Hippurate	nd (0%)	nd	nd	3 (13%)	nd	5
Myristate	5 (18%)	nd	14	25 (100%)	8	70
2-Oxoglutarate	2 (80%)	nd	9	7 (84%)	nd	23
Palmitoleate	2 (8%)	nd	6	31 (100%)	5	85
Palmitate	18 (93%)	nd	30	250 (100%)	75	780
Linoleate	11 (8%)	nd	14	110 (100%)	42	370
Oleate	36 (40%)	nd	120	460 (100%)	120	1830
Stearate	10 (87%)	nd	37	85 (100%)	31	470

The compounds are listed in the order of their chromatographic appearance, i.e. in the order of their MUs. The values given in parentheses after the means were the percentages of samples in which the compound could be detected and/or quantified (% Occ.). Abbreviations employed are Min. = minimal and Max. = maximal values. 'Trace' reflects values approximating 1 μmol/L, around the limit of detection. nd indicates not detectable, i.e. < 1 μmol/L

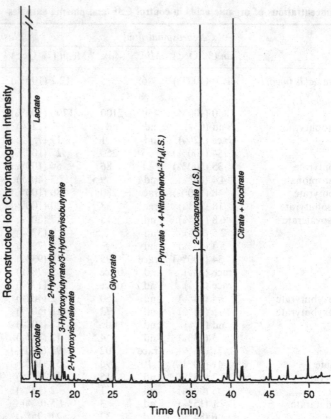

Figure 1 Reconstructed total-ion GCMS chromatogram of organic acids as pentafluoro-benzyloxime trimethylsilyl derivatives of 1 ml of control CSF

Table 2 CSF/plasma ratios of organic acids

	Mean	Min.	Max.
Total organic acids	0.8	0.4	2.4
Lactate	0.7	0.3	1.9
Glycolate	3.9	0.8	13
2- Hydroxybutyrate	0.8	0.3	1.8
3-Hydroxybutyrate	0.3	0.1	0.9
3- Hydroxyisobutyrate	1.2	0.5	2.9
Glycerate	5.9	1.2	22
Pyruvate	0.3	0.24	1.3
Pyroglutamate	1.1	0.5	0.7
2-Oxoisocaproate	0.1	0.03	0.2
Isocitrate	2.3	0.7	11
Citrate	2.6	0.7	12

Abbreviations employed are Min. = minimal and Max. = maximal values. The number of paired samples of CSF and plasma was 27

In general, concentrations of organic acids in CSF were slightly lower than in plasma, owing to the excretory function and the sink action of the CSF for cerebral metabolism. This was reflected in the total content of organic acids, which averaged 10.4 meq/L in CSF and 12.2 meq/L in plasma. However, there were substantial differences between different subgroups of organic acids. These differences are reflected in the ratios of individual acids between CSF and plasma (Table 2). Many compounds were present in CSF in substantially lesser amounts than in plasma. These comprised all of the metabolites related to fatty acid oxidation, acetoacetate and 3-hydroxybutyrate, long and medium straight-chain fatty acids as well as the branched-chain oxo acids and acids of the tricarboxylic acid cycle, such as succinate, malate, fumarate and 2-oxoglutarate. Citrate and isocitrate were found in higher concentrations in CSF than in plasma. As already discussed, the organic acid carrier system does not transfer di- or tricarboxylic acids (Oldendorf 1973). Long-chain fatty acids were not consistently detectable in CSF by this method; they showed a great variability and a pattern different from plasma and similar to previously reported data (Tichy and Skorkovska 1979), namely oleate ≥ palmitate ≥ stearate > linoleate > myristate > laurate.

Lactate and pyruvate were present in CSF in equal or slightly smaller amounts than in plasma. Certain organic acids in amino acid catabolic pathways, such as 2-hydroxybutyrate, 3-hydroxypropionate, 3-hydroxyisobutyrate, 2-hydroxyisovalerate and pyroglutamate had ratios approximating 0.7 to 1.2.

Some organic acids were found in substantially higher amounts in CSF than in plasma. These included glycolate, glycerate and 2,4-dihydroxybutyrate as well as citrate and isocitrate. Similar relations seemed to hold for erythronate and 2-hydroxyglutarate, but these could not be reliably quantitated in the majority of samples (Table 1). These observations are of special interest because some of these organic acids are known to be intermediates of neurotransmitter pathways, e.g. glycolate is a product of the catabolic pathway of 4-aminobutyrate (GABA) via 4-hydroxybutyrate (Gibson et al 1988). It can be anticipated that these compounds are also not transferred by the organic acid carrier system (Oldendorf 1973).

In addition to the metabolites listed in Table 1 a number of additional known as well as some unknown compounds were detected in substantial amounts in the paediatric control group, and subsequently in some samples of patients under investigation for unexplained neurological diseases, These included a number of different sugar acids as well as drug metabolites, e.g. caffeine and derivatives of valproic acid. In general, far fewer exogenous compounds interfere with organic acid analysis in CSF or plasma than in urine. Some metabolites were also regularly identified and quantified by this method for which better clinical chemical methods exist and/or for which reliable quantification would require a special handling of the samples, e.g. uric acid, ammonia, and cholesterol. These compounds were therefore not included in Table 1. It should, however, be kept in mind that the study of different sugar acids and uric acid in CSF may provide new details about cerebral metabolism in the future, as it has become apparent that some polyols are actively accumulated or synthesized in the CNS (Kusmierz et al 1989) and that the concentration of uric acid in the CSF may reflect alterations in nucleic acid catabolism (Fishman 1980).

In interpreting the data presented in Table 1 it must be remembered that, although care was taken to obtain reference samples from children who appeared to be as normal as possible, all suffered from illnesses that had been considered to require a diagnostic lumbar puncture. Presenting symptomatology included febrile convulsions, headaches, intermittent esotropia and paralysis of an extremity. In some instances anti-inflammatory, anticonvulsive or other drugs had been administered to the patients prior to lumbar puncture. It may be anticipated that the specimens utilized in this study reflect a greater variation of some metabolites than would have been found in completely healthy children receiving no medication. Greater variability of some organic acids was observed in the plasma samples of the paediatric controls in this study when compared to previously reported data on organic acids in plasma of 10 healthy adults (Hoffmann et al 1989). Plasma samples from the control children exhibited higher concentrations of metabolites related to fatty-acid oxidation, such as acetoacetate and 3-hydroxybutyrate and straight-chain fatty acids, and we also found some dicarboxylic acids that have not been detected in adults. There was great variability in the concentrations of these acids, which might reflect different degrees of catabolism or relatively prolonged fasting in moderately ill children. The reference ranges in Table 1, nevertheless, provide useful information, as samples to be investigated for metabolic disorders are also obtained from symptomatic patients and, more importantly, healthy children are generally not considered ethically reasonable subjects for lumbar puncture.

Considerations with regard to obtaining and handling the samples are relevant to the achievement of optimal analytical results. It appeared in this study that substantial loss of some oxo acids had occurred in some control samples. This was probably the result of inappropriate handling (delay in freezing) and prolonged storage of the samples at $-20°C$ instead of $-80°C$. Oxoacids were found in lower concentrations than in previous investigations (Hoffmann et al 1989). We have developed a standardized procedure for obtaining specimens of CSF for quantitative evaluation of metabolites. Samples are taken in the late morning 2 to 3 hours after breakfast, collected in 1 ml aliquots, and frozen on dry ice at the bedside. This allows the determination of a number of metabolites in addition to the organic acids, e.g. amino acids, GABA, pterins and folates. One sample vial should contain dithioerythritol/ diethylenetriaminepentaacetic acid as an antioxidant to permit investigation of pterins (Howells and Hyland 1987). The order in which the samples are taken is recorded, because of possible influences of the caudo-rostral gradient. Samples are stored at $-70°C$ or $-80°C$. At the same procedure a plasma sample of the patient is obtained, centrifuged at $4°C$, and stored at $-70°C$ or $-80°C$ until analysis.

ORGANIC ACIDS IN CSF IN ORGANOCIDOPOATHIES AND NEUROMETABOLIC DISORDERS

Until now data on organic acids in CSF in organoacidopathies have been very sparse despite the heightened awareness of the neurological manifestations in these disorders (Ozand and Gascon 1991). From the available data it appears that quantitative organic acid analysis in CSF will yield new information on the pathophysiology of these disorders and may prove necessary for the successful monitoring of treatment of organoacidopathies, which present mainly with neurological disease, e.g. *N*-acetylaspartic aciduria (Divry et al 1988), 'cerebral' lactic acidosis (Brown et al 1988; Benninger et al 1990), glutaryl-CoA dehydrogenase deficiency (Amir et al 1989; Hawarth et al 1991; Hoffmann et al 1991) and 4-hydroxybutyric aciduria (Gibson et al 1989; Jakobs et al 1990). Clinical presentation as primary or even exclusive neurological disease later in childhood or adulthood has also been reported in organoacidopathies, usually presenting in infancy with severe metabolic derangement, such as propionic acidaemia (Sethi et al 1989) and biotinidase deficiency (Fois et al 1986). It appears that analysis of organic acids in CSF should be included in the work-up and follow-up of such patients.

Concentrations of metabolites in CSF exceeding those of plasma may signify either a special role in cerebral metabolism or isolated deficiencies of CNS enzymes with (relatively) intact peripheral organ metabolism. An increase of methylmalonate in CSF out of proportion to plasma has been described in patients with cobalamin deficiency (Stabler et al 1991). In 4-hydroxybutyric aciduria the key metabolite 4-hydroxybutyrate, which is part of an important pathway in the nervous system as well as a known neuropharmacological agent (Mamelak 1989), is found increased in CSF out of proportion to plasma (Gibson et al 1989, 1990; Jakobs et al 1990). In non-ketotic hyperglycinaemia the elevated ratio of glycine in the CSF to that of plasma is currently the approach of choice to definitive diagnosis (Nyhan 1989). These observations highlight the need for quantitative analysis of metabolites for diagnosis (Fois et al 1986) as well as for monitoring of therapy (Fois et al 1986; Gibson et al 1989; Saijo et al 1991). Some recent studies in a number of patients with 'cerebral' organoacidopathies, which are summarized below, illustrate the potentials of this approach.

'CEREBRAL' LACTIC ACIDOSIS

Lactate elevation confined to CSF and brain has been described in biotinidase deficiency (Fois et al 1986) and in some mitochondriopathies (Brown et al 1988; Benninger et al 1990). Patients with 'cerebral' lactic acidosis show neurological symptoms, elevated levels of lactate and pyruvate in CSF, little or no systemic acidosis and levels of lactate, pyruvate and alanine in blood or urine so slightly elevated that they would be overlooked. Figure 2 shows a chromatogram of a CSF sample of a patient with cerebral lactic acidosis (lactate concentration 8.2 mmol/L). In plasma and urine, lactate concentrations fluctuated and were only modestly elevated. In addition, urinary 3-methylglutaconate was elevated. In CSF, no 3-methylglutaconate was found, but concentrations of 2-hydroxybutyrate (1700 μmol/L), 3-hydroxybutyrate (3600 μmol/L) and acetoacetate (200 μmol/L) were greatly elevated as were 3-hydroxyisovalerate (60 μmol/L) and 3-hydroxyvalerate (16 μmol/L). 5-Oxoproline (156 μmol/L), homovanillate (2 μmol/L) and other organic acids were also borderline elevated. In view of the relatively stable peripheral metabolism in this patient, we assume that some of the accumulation of organic acids in CSF is due to a competition at the blood–brain–CSF barrier in analogy with the effects of amino acid competition in the aminoacidopathies.

In other patients with mitochondriopathies, we found lesser elevations of lactate

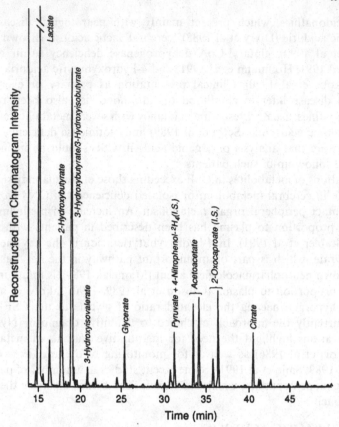

Figure 2 Reconstructed total-ion GCMS chromatogram of organic acids as pentafluoro-benzyloxime trimethylsilyl derivatives of 1 ml of CSF from a patient with lactic acidosis and 3-methylglutaconic aciduria

and pyruvate in CSF — despite higher concentrations of plasma lactate. In one of these patients significant amounts of 2,3-butanediol ($\approx 200\,\mu mol/L$) were noted in CSF in addition to elevated lactate as an isolated finding with no explanation so far. These observations are still preliminary. However, we think that repeated analyses of organic acids in CSF is the best way to monitor treatment in patients with 'cerebral' lactic acidosis. Additional information relevant to the pathophysiology and the metabolic derangements in the single patient may be obtained by general organic acid analysis, which will be missed by enzymatic determinations of lactate.

DISORDERS OF PROPIONATE AND METHYLMALONATE METABOLISM

Progressive neurological deterioration, primarily extrapyramidal, with acute episodic vomiting and lethargy, are important sequelae of propionic acidaemia and disorders of methylmalonate metabolism. NMR imaging of the CNS has shown bilateral pallidal necrosis in the majority of patients with methylmalonic acidaemia, both in those acutely unwell and in those without such episodes (Andreula et al 1991). There is evidence that the accumulated metabolites are specifically cytotoxic to the globus pallidus (Dayan and Ramsey 1984), but there is also general inhibition of mitochondrial respiration by toxic metabolites in both methylmalonic and propionic acidaemia (Hayasaka et al 1982).

We found total organic acids in CSF samples from patients with methylmalonic and propionic acidaemia to be highly elevated (35–70 meq/L). The pattern of organic acids in CSF in methylmalonic acidaemia was quite different from that in propionic acidaemia. In good metabolic control, an elevated methylmalonate was the predominant finding in methylmalonic acidaemia, often without other obvious abnormalities (Table 3). In patients with inborn errors of adenosyl and methyl cobalamin, methylmalonate became normal on daily hydroxy-B_{12}. Stabler and colleagues (1991) recently determined methylmalonate levels in CSF and plasma from controls and found concentrations of methylmalonate in CSF always higher than in plasma (Table 3). In patients with neuropsychiatric syndromes due to cobalamin deficiency, methylmalonate in CSF was further increased out of proportion to plasma values. These findings were interpreted as an indication 'that this metabolic pathway may be important in the normal functioning of the nervous system' and 'that impairment of this pathway may play a role in the neurological damage caused by lack of cobalamin.'

In CSF samples from patients with propionic acidaemia we found elevations of 3-hydroxypropionate (up to 150 μmol/L), 3-hydroxyisovalerate (up to 150 μmol/L) and methylcitrate (up to 20 μmol/L) with little variation in response to good or poor metabolic control (Figure 3). Increased concentrations were also demonstrated for lactate (up to 4500 μmol/L), pyruvate (up to 400 μmol/L), 2-hydroxybutyrate (up to 400 μmol/L), 3-hydroxybutyrate (up to 1600 μmol/L), and acetoacetate (up to 400 μmol/L). The branched-chain oxo acids were also borderline elevated.

Studies of metabolites in CSF may contribute to a better understanding of the pathomechanisms of neurological damage in disorders of propionate and methylmalonate metabolism. In one child with propionic acidaemia, who was in good metabolic control but suffered a sudden bilateral infarction of the basal ganglia, total organic acids were 51 meq/L in CSF and 25 meq/L in plasma at the time of the cerebral insult. Methylcitrate was four times higher in CSF than in plasma. Concentrations of 3-hydroxypropionate and 3-hydroxyisovalerate were similar in CSF and plasma, while propionylglycine was only detectable in plasma (20–40 μmol/L). Lactate and pyruvate showed higher elevations in CSF than in plasma. Plasma citrate concentration was abnormally low, plasma isocitrate was undetectable, and ketone bodies were low. From our current knowledge it is difficult to interpret these data conclusively, but they indicate a particular derangement of brain metabolism and illustrate the need for a better understanding of the consequences of the enzyme defects in methylmalonic and propionic acidaemia in the brain. This knowledge will be necessary to improve treatment regimens to prevent or at least reduce neurological damage.

Table 3 Levels of methylmalonate in CSF and plasma samples in different disorders

Diagnosis	Cerebrospinal fluid (μmol/L)			Plasma (μmol/L)			CSF/plasma		
	Mean	Min.	Max.	Mean	Min.	Max.	Mean	Min.	Max.
Controls (n = 58)[a]	0.33	0.14	0.73	0.14	0.04	0.26	2.7	1.2	7.8
Cobalamin deficiency (n = 4)[a]	191	15	358	22	4.3	37	8.4	3.5	13.5
Methylmalonic acidaemia: Mut[0] (n = 3)	200	121	342	—	—	—	—		
Cobalamin mutants C & D (n = 2)	nd								

[a]Data were obtained with a stable isotope dilution selective-ion monitoring technique by Stabler et al (1991). Abbreviations employed are Min. = minimal and Max. = maximal values. nd indicates not detectable, i.e. < 1 μmol/L, — = undetermined

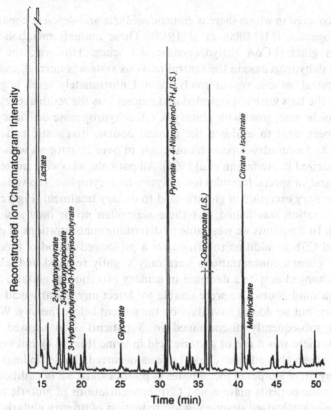

Figure 3 Reconstructed total-ion GCMS chromatogram of organic acids as pentafluorobenzyloxime trimethylsilyl derivatives of 1 ml of CSF from a patient with L-propionic acidaemia

GLUTARYL-CoA DEHYDROGENASE DEFICIENCY

Glutaryl-CoA dehydrogenase deficiency presents as an acute to subacute encephalopathy of infancy or early childhood. Because of the lack of significant metabolic derangements or crises, the disease appears to remain frequently undiagnosed or misdiagnosed (Amir et al 1989; Haworth et al 1991; Hoffmann et al 1991). The mechanisms leading to the neuropathological changes are still unclear. The predominant effects on the basal ganglia appear to be specific, as neurons of the caudate nuclei and the putamen are especially sensitive to glutaric acid. Glutaric, glutaconic and 3-hydroxyglutaric acids inhibit glutamic acid decarboxylase activity in the basal ganglia and consequently the production of GABA (Stokke et al 1976). In two patients glutaric acid was determined in the brain after death and very high levels, around 1 μmol/g wet weight were found (Goodman 1977; Leibel et al 1980), close to the K_i for glutamic acid decarboxylase in rat and rabbit brain (Stokke et al 1976). In these patients, neurons were lost and glutamic acid decarboxylase activity and GABA content in the basal ganglia were profoundly reduced (Goodman 1977; Leibel et al 1980). An experiment of nature should also be mentioned. Fruit-eating bats were discovered in whom there is glutaric aciduria and deficient hepatic glutaryl-CoA dehydrogenase (McMillan et al 1988). These animals metabolically mimic patients with glutaryl-CoA dehydrogenase deficiency. However, the activity of glutaryl-CoA dehydrogenase in the central nervous system is normal, and they show no signs of central nervous system involvement. Unfortunately, levels of glutaric acid in the CSF of the bats were not reported and neither was the residual enzyme activity in the brain of homozygotes with glutaryl-CoA dehydrogenase deficiency.

We have been able to evaluate the clinical course, diagnostic and therapeutic management and neurodevelopmental outcome in over 20 patients with this disease (partly summarized in Hoffmann et al 1991). All patients, who were treated with low-protein diets and/or special formulas low in lysine and trytophan, showed a favourable response of urinary excretion of glutaric acid to dietary treatment (Figure 4). Further clinical deterioration was halted, but there was often no or only a slight clinical improvement. In 5 patients we were able to determine concentrations of glutaric acid in plasma and CSF in addition to urine over a prolonged period of time (Figure 4). In 4 patients plasma concentrations were only slightly reduced and levels in CSF remained unchanged despite a decrease in urinary excretion of glutaric acid. In one patient with a mild course, we were unable to detect any glutaric acid in the CSF during therapy, but we do not have data on the patient before therapy. When dietary therapy was subsequently discontinued in 3 patients who showed no clinical improvement, there was a rise of glutaric acid in urine (Figure 4), but not in plasma or CSF. No observable adverse clinical effects occurred. These findings need to be followed up by more studies of metabolites in plasma and CSF in addition to urinary analyses. In some patients plasma and CSF concentrations of glutaric acid seem to remain virtually unchanged despite a good reduction of urinary glutaric acid; and it is probable that in these patients strict dietary therapy can be relaxed or even discontinued. Carnitine substitution must, however, be continued to prevent further neurometabolic crises. Other patients are making slow but definitive progress and should remain on dietary therapy.

Considering the severity of the disease it appears justifiable to perform lumbar punctures every 3–6 months in patients with glutaryl-CoA dehydrogenase deficiency to evaluate the effect of therapy. We hope that this approach will give us some indications on the efficiency of dietary treatment in the single patient and will provide some information about the pathogenetic mechanisms of the disease. Clinical evaluation of therapy may be inconclusive for a long time, because part of the neurological deficit in symptomatic patients will be due to permanent structural damage as a result of the initial encephalopathic crisis; this can never be improved by subsequent therapeutic regimens. In our experience the urinary excretion of glutaric acid appears to be an inappropriate parameter to follow treatment, at least without plasma and CSF determinations. This is also true when bound urinary glutaric acid is measured in addition to free glutaric acid in the urine (Ribes et al 1992). This can be a useful aid in some difficult cases for the diagnosis of glutaryl-CoA dehydrogenase deficiency, but for the monitoring of treatment it suffers from the same limitations as determinations of free glutaric acid in the urine alone.

2-HYDROXYGLUTARIC ACIDURIA

2-Hydroxyglutaric acid is another example of a 'cerebral' organic acid disorder which in the scanty reported clinical since its description has been detected primarily. Barth et al. (1993) and the additional patient has shown a round-the-clock 6-month in urine and plasma in the pathology by several cases. In one case of these two patients, a poor anticonvulsant effect (Hoffmann et al. 1990, Barth et al. 1992). The metabolic picture of the disorder was elucidated with research into highly elevated concentrations of 2-hydroxyglutaric acid in urine, plasma and CSF. High concentrations in plasma and also in the CSF level, and no other abnormalities were found of organic acid apart from the 2-hydroxyglutaric acid. We studied CSF and plasma samples from two siblings with 2-hydroxyglutaric aciduria. A chromatogram of CSF sample is shown in Figure 3. The patients were initially treated by various dietary or organic acids which reduced only very restricted light-elevated concentrations of 2-hydroxyglutaric acid ranged in plasma 3000 μmol/mol creatinine; in serum in front of the controls of these 2-hydroxyglutaric acidaemia. (Hoffmann...

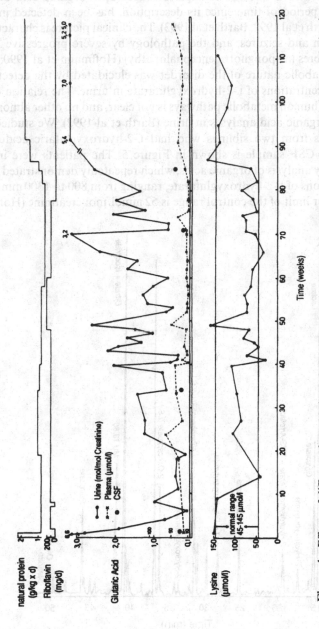

Figure 4 Effects of different therapeutic regimens on levels of glutaric acid in urine, plasma and CSF, and lysine in plasma in a patient with glutaryl-CoA dehydrogenase deficiency. Concentrations of glutaric acid in CSF are also reported in μmol/L

L-2-HYDROXYGLUTARIC ACIDURIA

L-2-Hydroxyglutaric acid is another example of a 'cerebral' organic acid disorder which, in the short period of time since its description, has been detected in more than 10 patients (Barth et al 1992; Barth et al 1993). The clinical picture is characterized by mental regression and seizures, and the pathology by severe progressive loss of myelinate arcuate fibres (a spongiform encephalopathy) (Hoffmann et al 1990; Barth et al 1992). The metabolic nature of the disorder was elucidated by the detection of highly elevated concentrations of L-2-hydroxyglutarate in urine. The relation of L-2-hydroxyglutarate to human metabolic pathways is not clear, and no other abnormalities were found on organic acid analyses in urine (Barth et al 1992). We studied CSF and plasma samples from two siblings who had L-2-hydroxyglutaric aciduria. A chromatogram of a CSF sample is shown in Figure 5. The patients were initially diagnosed by urinary analysis of organic acids, which repeatedly demonstrated highly elevated concentrations of L-2-hydroxyglutarate, ranging from 800 to 1300 mmol/mol creatinine. The upper limit of the control range is 52 mmol/mol creatinine (Hoffmann

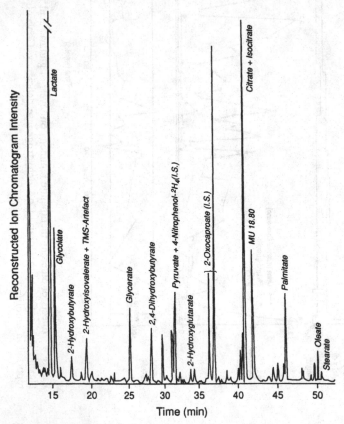

Figure 5 Reconstructed total-ion GCMS chromatogram of organic acids as pentafluorobenzyloxime trimethylsilyl derivatives of 1 ml of CSF from a patient with L-2-hydroxyglutaric aciduria

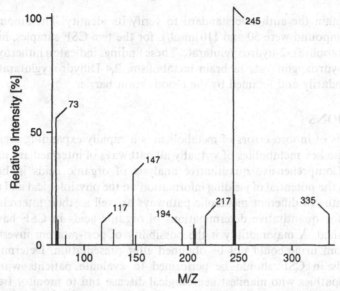

Figure 6 Spectrum of unknown compound (presumably 2,4-dihydroxyglutarate) present in large amounts in CSF in both patients with L-2-hydroxyglutaric aciduria

et al 1989). There were no other abnormalities detectable on urinary organic acid analysis. Quantitative study of CSF and plasma organic acids revealed elevated concentrations of 2-hydroxyglutarate. The concentrations of 2-hydroxyglutarate in CSF were 4.4 to 5.8 times the plasma levels. Concentrations of glycolate, glycerate, 2,4-dihydroxybutyrate, citrate and isocitrate were also elevated in the CSF, although levels were normal in plasma. These were the same acids that in control CSF were found in higher concentrations than in plasma (Tables 1 and 2). Some of these organic acids are known to be relevant to neurotransmitter metabolism. Glycolate is related to the degradation of GABA via 4-hydroxybutyrate (Gibson et al 1988). Fatty acids were also found to be elevated in CSF, suggesting an increased breakdown of lipids in the CNS. Lactate was slightly elevated in CSF of both patients and in the plasma of one. In addition, an unknown compound, which has not previously been encountered in substantial amounts in any physiological fluid, was found in the CSF of both patients (Figure 6). This compound was characterized by its methylene unit of 18.80 on a DB5 wide-bore fused-silica capillary column. A computer search gave 2,4-dihydroxyglutarate as the most likely candidate. Unfortunately, we are as yet

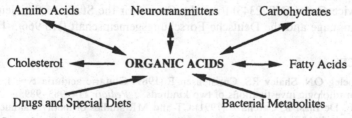

Figure 7 Organic acids in intermediary metabolism

unable to obtain the authentic standard to verify its identity. The amounts of the unknown compound were 50 and 110 μmol/L for the two CSF samples, higher than the index metabolite L-2-hydroxyglutarate. These findings indicate a hitherto unknown role of L-2-hydroxyglutarate in brain metabolism. 2,4-Dihydroxyglutarate may be formed secondarily and retained by the blood–brain barrier.

CONCLUSIONS

The diagnosis of inborn errors of metabolism is a rapidly expanding area. Organic acids comprise key metabolites of virtually all pathways of intermediary metabolism (Figure 7). Comprehensive quantitative analysis of organic acids in body fluids therefore has the potential of yielding information on the physiological and pathophysiological status of different metabolic pathways as well as their interrelationships. Indications for quantitative determinations of organic acids in CSF have not yet been established. A major utility is the possibility of post-mortem investigation in those in whom urine could not be obtained after presentation. Determinations of organic acids in CSF should be performed to evaluate patients with different organoacidopathies who manifest neurological disease and to monitor treatment of 'cerebral' organoacidopathies such as 'cerebral' lactic acidosis, disorders of propionate and methylmalonate metabolism and glutaryl-CoA dehydrogenase deficiency.

Investigations of metabolites in CSF or selective screening of CSF for neurometabolic diseases is a relatively new and developing area of patient care and research. Progressive or static encephalopathies, myelopathy-neuropathy, myopathy and even brain dysgenesis are increasingly being recognized as important aspects or even the leading symptom of many patients with organoacidopathies (Ozand and Gascon 1991). Metabolic derangements such as acidosis or hypoglycaemia may be absent altogether (Sethi et al 1989; Hoffmann et al 1991; Ozand and Gascon 1991; Barth et al 1992). It is likely that there are additional disorders yet to be defined. These could be called 'cerebral' organic acid disorders, and some may only be identifiable by quantitative analysis of organic acids in CSF.

ACKNOWLEDGEMENTS

We thank Dr H. J. Christen and Dr F. Hanefeld, Göttingen, Dr H.-P. Hartung and Dr G. Reimann, Würzburg, Dr B. Lawrenz-Wolf, Kassel, and Dr E. Pronicka, Warsaw, for providing samples of their patients. Excellent technical assistance was provided by Claudia K. Seppel, Bonnie Holmes, Chris LaRosa, Brigitte Schmidt-Mader and Laverne Mitchell. This study was in part supported by the US Public Health Service Grant (NS22343) from the Center for the Study of the Neurological Basis of Language and the Deutsche Forschungsgemeinschaft (Ho 966/3-1).

REFERENCES

Amir N, Elpeleg ON, Shalev RS, Christensen E (1989) Glutaric aciduria type I: enzymatic and neuroradiologic investigations of two kindreds. *J Pediatr* **114**: 983–989.
Andreula CF, De Blasi R, Carella A (1991) CT and MR studies of methylmalonic acidemia. *Am J Neuroradiol* **12**: 410–412.

Barth PG, Hoffmann GF, Jaeken J et al (1992) L-2-Hydroxyglutaric acidemia: A novel inherited neurometabolic disease. *Ann Neurol* 32: 66–71.

Barth PG, Hoffmann GF, Jaeken J et al (1993) L-2-Hydroxyglutaric acidaemia: clinical and biochemical findings in 12 patients and preliminary report on L-2-hydroxyacid dehydrogenase. *J Inher Metab Dis* 16: 753–761.

Benninger C, Lichter-Konecki U, Schmitt HP, Reichmann H (1990) Isolierter Cytochrom-C-Oxidase-Mangel im Gehirn. In Hanefeld, F, Rating D, Christen HJ, eds. *Aktuelle Neuropädiatrie* 1989. Heidelberg: Springer Verlag, 217–220.

Brown GK, Haan EH, Kirby DM et al (1988) 'Cerebral' lactic acidosis: defects in pyruvate metabolism with profound brain damage and minimal systemic acidosis. *Eur J Pediatr* 147: 10–14.

Chalmers RA, Lawson AM (1982) *Organic Acids In Man.* London: Chapman and Hall.

Chang Y-F (1976) Pipecolic acid pathway: The major lysine metabolic route in the rat brain. *Biochem Biophys Res Commun* 69: 174–180.

Coude M, Kamoun P (1992) Organic acids in post-mortem cerebrospinal fluid. *Clin Chim Acta* 206: 201–206.

Coude M, Chadefaux B, Charpentier C, Kamoun P (1991) Medium-chain triglycerides: a pitfall in the assay of organic acids in cerebrospinal fluid. *J Inher Metab Dis* 14: 841.

Cremer JE, Braun LD, Oldendorf WH (1976) Changes during development in transport processes of the blood–brain barrier. *Biochim Biophys Acta* 448: 633–637.

Cutler RWP, Page L, Galicich J, Watters GV (1968) Formation and absorption of cerebrospinal fluid in man. *Brain* 91: 707–720.

Dayan AD, Ramsey RB (1984) An inborn error of vitamin B_{12} metabolism associated with cellular deficiency of coenzyme forms of the vitamin. *J Neurol Sci* 23: 117–128.

Divry P, Vianey-Liaud C, Gay C, Macabeo V, Rapin F, Echenne B (1988). *N*-acetylaspartic aciduria: report of three new cases in children with a neurological syndrome associating macrocephaly and leukodystrophy. *J Inher Metab Dis* 11: 307–308.

Divry P, Vianey-Liaud C, Jakobs C, Ten-Brink HJ, Dutruge J, Gilly R (1990) Sudden infant death syndrome: organic acid profiles in cerebrospinal fluid from 47 children and the occurrence of *N*-acetylaspartic acid. *J Inher Metab Dis* 13: 330–332.

Fenstermacher JD, Patlak CS (1975) The exchange of material between cerebrospinal fluid and brain. In Cserr HF, Fenstermacher JD, Fencl V, eds. *Fluid Environment of the Brain.* New York: Academic Press, 201–214.

Fishman RA (1980) *Cerebrospinal Fluid in Diseases of the Nervous System.* Philadelphia: WB Saunders.

Fois A, Cioni M, Balestri P, Bartilini G, Baumgartner R, Bachmann C (1986) Biotinidase deficiency: metabolites in CSF. *J Inher Metab Dis* 9: 284–285.

Gerrits GP, Trijbels FJ, Monnens LA et al (1989) Reference values for amino acids in cerebrospinal fluid of children determined using ion-exchange chromatography with fluorimetric detection. *Clin Chim Acta* 182: 271–280.

Gibson KM, Hoffmann G, Sweetman L et al (1988) 4-Hydroxybutyric aciduria in a patient without ataxia or convulsions. *Eur J Pediatr* 147: 529–531.

Gibson KM, DeVivo DC, Jakobs C (1989) Vigabatrin therapy in a patient with succinic semialdehyde dehydrogenase deficiency. *Lancet* 2: 1105–1106.

Gibson KM, Aramaki S, Sweetman L et al (1990) Stable isotope dilution analysis of 4-hydroxybutyric acid: an accurate method for quantification in physiological fluids and the prenatal diagnosis of 4-hydroxybutyric aciduria. *Biomed Environ Mass Spectrom* 19: 89–93.

Goodman SI, Norenberg MD, Shikes RH, Breslich DJ, Moe PG (1977) Glutaric aciduria: biochemical and morphological considerations. *J Pediatr* 90: 746–750.

Haworth JC, Booth FA, Chudley AE et al (1991) Phenotypic variability in glutaric aciduria type I: Report of fourteen cases in five Canadian Indian kindreds. *J Pediatr* 118: 52–58.

Hayasaka K, Metoki K, Satoh T, Narisawa K, Tada K, Kawakami T (1982) Comparison of cytosolic and mitochondrial enzyme alteration in the livers of propionic or methylmalonic acidemia: a reduction of cytochrome oxidase activity. *Tohoku J Exp Med* 137: 329–334.

Hoffmann G, Sweetman L (1987) O-(2,3,4,5,6-Pentafluorobenzyl)oxime-trimethylsilyl ester derivatives for quantitative gas chromatographic–mass spectrometric studies of aldehydes, ketones, and oxoacids in biological fluids. *J Chromatogr* **421**: 336–343.

Hoffmann GF, Sweetman L (1991) O-(2,3,4,5,6-Pentafluorobenzyl)oxime-trimethylsilyl ester derivatives for sensitive identification and quantitation of aldehydes, ketones, and oxoacids in biological fluids. *Clin Chim Acta* **199**: 237–242.

Hoffmann G, Aramaki S, Blum-Hoffmann E, Nyhan WL, Sweetman L (1989) Quantitative analysis for organic acids in biological samples: batch isolation followed by gas chromatographic–mass spectrometric analysis. *Clin Chem* **35**: 587–595.

Hoffmann GF, Hunneman DH, Voss W et al (1990) L-2-Hydroxyglutarazidurie: Eine neue Enzephalopathie mit leukodystrophen Veränderungen. In Hanefeld F, Rating D, Christen HJ, eds. *Aktuelle Neuropädiatrie 1989*. Heidelberg: Springer Verlag, 139–142.

Hoffmann GF, Trefz FK, Barth P et al (1991) Glutaryl-CoA dehydrogenase deficiency: A distinct encephalopathy. *Pediatrics* **88**: 1194–1203.

Howells DW, Hyland K (1987) Direct analysis of tetrahydrobiopterin in cerebrospinal fluid by high-performance liquid chromatography with redox electrochemistry – Prevention of autooxidation during storage and analysis. *Clin Chim Acta* **167**: 23–30.

Jakobs C, Smit LME, Kneer J, Michael T, Gibson KM (1990) The first adult case with 4-hydroxybutyric aciduria. *J Inher Metab Dis* **13**: 341–344.

Knudsen GM, Paulson OB, Hetz MM (1991) Kinetic analysis of the human blood–brain barrier transport of lactate and its influence by hypercapnia. *J Cereb Blood Flow Metab* **10**: 698–703.

Kusmierz J, DeGeorge JJ, Sweeney D, May C, Rapoport SI (1989) Quantitative analysis of polyols in human plasma and cerebrospinal fluid. *J Chromatogr* **497**: 39–48.

Leibel RL, Shih VE, Goodman SI et al (1980) Glutaric acidemia: a metabolic disorder causing progressive choreoathetosis. *Neurology* **30**: 1163–1168.

Mamelak M (1989) Gammahydroxybutyrate: An endogenous regulator of energy metabolism. *Neurosci Biobehav Rev* **13**: 187–198.

McMillan TA, Gibson KM, Sweetman L, Meyers GS, Green R (1988) Conversation of central nervous system glutaryl-coenzyme A dehydrogenase in fruit-eating bats with glutaric aciduria and deficient hepatic glutaryl-CoA dehydrogenase. *J Biol Chem* **263**: 17258–17261.

Müller J, Vahar-Martiar H (1975) Freie Fettsäuren und Kohlenwasserstoffe im Liquor cerebrospinalis. *Z Klin Chem Klin Biochem* **13**: 183–185.

Nyhan WL (1989) Nonketotic hyperglycinemia. In Scriver CR, Beaudet AL, Sly WS, Valle D, eds. *The Metabolic Basis of Inherited Disease*, 6th edn. New York: McGraw-Hill, 743–753.

Oldendorf WH (1973) Carrier-mediated blood–brain barrier transport of short-chain monocarboxylic organic acids. *Am J Physiol* **224**: 1450–1453.

Ozand PT, Gascon GG (1991) Organic acidurias: A review. Part 1. *J Child Neurol* **6**: 196–219.

Pardridge WM, Connor JD, Crawford IL (1975) Permeability changes in the blood–brain barrier: Causes and consequences. *CRC Crit Rev Toxicol* **3**: 159–199.

Rall DP, Stabenau JR, Zubrod CG (1959) Distribution of drugs between blood and cerebrospinal fluid: general methodology and effect of pH gradients. *J Pharmac Exp Ther* **125**: 185–193.

Rapoport SI (1976) *Blood–Brain Barrier in Physiology and Medicine*. New York: Raven Press.

Ribes A, Riudor E, Briones P, Christensen E, Campistol J, Millington DS (1992) Significance of bound glutarate in the diagnosis of glutaric aciduria type I. *J Inher Metab Dis* **15**: 367–370.

Roe CR, Gale DS, Millington DS (1988) Evidence for primary CNS 'organic acidurias'. *SSIEM, Proceedings of the 26th Annual Symposium, Glasgow, Sept. 1988*: P68 (abstract).

Saijo T, Naito E, Ito M, Takeda E, Hashimoto T, Kuroda Y (1991) Therapeutic effect of sodium dichloroacetate on visual and auditory hallucinations in a patient with MELAS. *Neuropediatrics* **22**: 166–167.

Sethi KD, Ray R, Roesel RA et al (1989) Adult-onset chorea and dementia with propionic acidemia. *Neurology* **39**: 1343–1345.

Stabler SP, Allen RH, Barrett RE, Savage DG, Lindenbaum J (1991) Cerebrospinal fluid methylmalonic acid levels in normal subjects and patients with cobalamin deficiency. *Neurology* **41**: 1627–1632.

Stokke O, Goodman SI, Moe PG (1976) Inhibition of brain glutamate decarboxylase by glutarate, glutaconate, and beta-hydroxyglutarate: explanation of the symptoms in glutaric aciduria? *Clin Chim Acta* **66**: 411–415.

Sweetman L (1991) Organic acid analysis. In Hommes FA, ed. *Techniques in Diagnostic Human Biochemical Laboratories: A Laboratory Manual*. New York: Wiley-Liss, 143–176.

Tichy J, Skorkovska I (1979) Spectrum of total fatty acids in cerebrospinal fluid determined by gas chromatography. *J Chromatogr* **162**: 185–196.

Weyne J, Leusen I (1975) Lactate in CSF in relation to brain and blood. In Cserr HF, Fenstermacher JD, Fencl V, eds. *Fluid Environment of the Brain*. New York: Academic Press, 255–276.

J. Inher. Metab. Dis. 16 (1993) 670–675

Cerebrospinal Fluid Amino Acids, Purines and Pyrimidines as a Tool in the Study of Metabolic Brain Diseases

G. P. J. M. Gerrits, L. A. H. Monnens, F. J. M. Gabreëls, R. A. De Abreu, A. Koster and J. M. F. Trijbels
Departments of Pediatrics, Neurology and Statistical Consultation, University of Nijmegen, P.O. Box 9101, 6500 HB Nijmegen, The Netherlands

Summary: After establishing more extended reference values for amino acids, purines and pyrimidines in cerebrospinal fluid (CSF) in infancy and childhood, we studied 1250 CSF-aliquots from patients who were undergoing a diagnostic lumbar puncture for diverse clinical indications. Our primary aim was to answer the question whether determination of the concentration of amino acids, purines and pyrimidines in CSF is a useful tool in screening for metabolic disorders in children with unexplained mental retardation.

In unexplained mental retardation (95 patients) we observed varying abnormalities of CSF. These were reproducible in only 2 patients (a decrease of homocarnosine in combination with two unidentified compounds). Striking abnormalities in pyrimidine content which are limited to CSF are found in argininosuccinic aciduria and uraemia. In uraemia a general decrease in amino acids in CSF and increase of γ-aminobutyric acid (GABA) was observed.

The results obtained indicate that determination of amino acids, purines and pyrimidines in CSF is only of limited value in the diagnosis of unexplained mental retardation.

INTRODUCTION

Metabolic brain diseases can be associated with abnormal cerebrospinal fluid (CSF) levels of amino acids, purines and pyrimidines. In some disorders, like Leigh's syndrome (Van Erven et al 1987) and non-ketotic hyperglycinaemia (Nyhan 1989), the presence in CSF of high concentrations of specific metabolites (lactate and glycine, respectively) is a diagnostic feature. In argininosuccinic aciduria, a urea cycle defect, the concentration of argininosuccinic acid and its anhydrides in CSF is found to be twice to four times that noted in plasma (Scriver and Rosenberg 1973; McKusick 20790). Furthermore, it is evident that neurotransmitter metabolism shows abnormalities limited to CSF, e.g. disorders in γ-aminobutyric acid (GABA) metabolism are reported with either increased or decreased CSF levels of GABA (Jaeken et al 1990).

The availability of new and more sensitive methods for determining amino acids, purines and pyrimidines urged us to establish more extended reference values of these compounds in CSF (Gerrits et al 1988; Gerrits et al 1989). Our aim was to study CSF amino acid, purine and pyrimidine levels in various neurological disorders in order to contribute to the aetiological unravelling of metabolic brain diseases.

We studied amino acids, purines and pyrimidines in CSF of a large number of CSF-aliquots (1250) from patients who were undergoing a diagnostic lumbar puncture for diverse clinical indications.

MATERIALS AND METHODS

Samples and patients

Specimens of CSF (1250) were obtained from 1154 different subjects ranging in age from 0–18 years, who were undergoing a diagnostic lumbar puncture for conventional clinical indications such as suspected central nervous system infection or other neurological disorders. Whenever possible, 1 ml of each CSF sample was put aside for the present investigation. The CSF samples were kept frozen at $-70°C$ until analysis. Just before analysis the samples were thawed and deproteinized according to previous published methods (Gerrits et al 1989).

The subjects were divided into different clinical diagnostic groups by retrospective study of the patient records (see Table 1). The total number of patients studied is 856, because we had to exclude retrospectively a number of patients owing to too high erythrocyte counts in CSF or technical problems concerning the amount of CSF. Some patients had more than one diagnosis. Ninety-five patients were selected with mental retardation without any other diagnosis. The values obtained were compared with previously established normal values of amino acids, purines and pyrimidines (Gerrits et al 1988; Gerrits et al 1989). Our reference group has been extended compared to our earlier publications; the reference ranges have been confirmed and remain the same.

In addition the coefficient of variability was calculated from repeated sampling of CSF from a group of 13 patients with stable leukaemia without central nervous system localization (see Table 2).

Table 1 The diagnostic groups in the patients (*n* = number of patients)

	n
1. Reference group normal children	122
2. Rest group	265
3. Diagnostic groups:	
metabolic disorders	27
heredodegenerative disorders	31
structural congenital disorders of the brain	27
perinatal pathology	49
febrile convulsions	32
epilepsy	134
unexplained mental retardation	180
mental retardation with a known cause	35
syndromes	17
infections such as meningitis	36
oncology such as leukaemia	134

Table 2 The coefficient of variability for the different compounds in CSF ($n = 13$)

taurine	10.5	leucine	9.8
phosphoethanolamine	11.6	tyrosine	15.8
aspartic acid	22.2	phenylalanine	6.9
threonine	18.8	γ-aminobutyric acid	28.0
serine	12.3	ethanolamine	12.1
asparagine	11.7	tryptophan	27.2
α-aminoadipic acid	30.2	ornithine	19.1
glycine	17.5	lysine	12.8
alanine	13.5	histidine	14.4
citrulline	13.2	N-ε-methyl-l-lysine	18.4
α-aminobutyric acid	20.2	3-methylhistidine	22.1
valine	10.1	homocarnosine	14.9
methionine	20.2	arginine	15.4
cysthathionine	23.7		
isoleucine	11.4		

Analyses

For detailed information about the chemical analyses the reader is referred to previous publications (Gerrits et al 1988; Gerrits et al 1989).

Statistics

The probability that a given number of subjects falls outside the reference values (defined as the interval between the 2.5 and 97.5 centile) is calculated using a binomial distribution. The significance level is set as 0.01.

RESULTS

Three groups showed abnormalities in either amino acids and/or purines and pyrimidines.

From the group of 'metabolic diseases' we found abnormalities in three patients with arginosuccinic aciduria. Besides the known deviations in amino acids (increases in argininosuccinic acid, its anhydrides and citrulline) we found an elevation in CSF of pseudouridine and uridine, which was not found in urine and plasma (Gerrits et al 1993).

The second group contains ten patients, aged 2–59 months, with chronic renal failure and a disturbed mental development. All patients showed increased concentrations of pseudouridine in CSF up to 1000 times the normal value. Plasma showed concentrations up to 10 times the normal value. Eight out of ten showed a cytidine increase limited to CSF. Plasma values were normal or slightly increased. Cytidine in CSF was even greater than the respective plasma values in eight out of ten patients. Also a large peak of an unknown compound was detected in the pyrimidine-spectrum of each patient (Gerrits et al 1991).

In the patients with chronic renal failure a significant decrease in CSF was observed for taurine, phosphoethanolamine, aspartic acid, serine, alanine, α-aminobutyric acid, valine, isoleucine, leucine, tyrosine, tryptophan, lysine, histidine, N-ε-methyl-l-lysine and homocarnosine ($p \leq 0.001$). A significant increase of GABA and 3-methylhistidine was observed (both $p < 0.001$). Plasma citrulline, α-aminobutyric acid, tryptophan and cystine were significantly decreased (p-values < 0.001) while glycine concentration was increased ($p = 0.004$).

In three of these patients the amino acid concentrations in CSF were determined again after more than six weeks of continuous peritoneal ambulatory dialysis (CAPD) treatment. Similar results were obtained as before CAPD treatment, only 3-methylhistidine normalized.

The largest group in our study, concerns the group with mental retardation of unknown aetiology. In this group of 95 patients careful clinical selection was done to obtain a group with no other diagnosis. Within this group, 52 patients showed deviations of more than 2SD in one or more CSF amino acids, and from these 19 patients showed deviations of more than 3SD. Fifteen of this last group of 19 were asked for permission for a repeated lumbar puncture in order to investigate whether the observed abnormalities were reproducible. Nine patients agreed. In the CSF of of these patients we could not confirm the abnormalities, except for two patients. They showed a decrease of homocarnosine and the presence of two unknown peaks in CSF. GABA was normal. In both patients, the two hitherto unidentified peaks in the CSF-amino acid pattern were not found in serum.

Thirty-two of the 95 retarded patients showed deviations of more than 2SD in one or more CSF purine and pyrimidine concentrations. From these we selected two groups: one group of 11 patients with 5SD deviations in one or more purine or pyrimidine but no typical pattern, and a second group of 22 patients with, amongst other purine and pyrimidine deviations, a cytidine elevation of more than 2SD. One patient appeared in both groups. The last group was selected because we noted that cytidine concentration was abnormal in more cases than for the other purine and pyrimidine compounds. From the first group 8 patients responded to our request for a further lumbar puncture. None of them showed reproducible patterns in CSF purines and pyrimidines. Of the group with the elevated cytidine concentration 12 responded; in no patient was the cytidine elevation reproducible.

DISCUSSION

Reference values for amino acids in CSF of young infants and children have been infrequently reported (Heiblim et al 1978; Applegarth et al 1979; Honda 1984). As far as reported, these data have been collected from measurements using ion-exchange chromatography with the ninhydrin detection system. A few investigators have used HPLC procedures with fluorimetric detection to characterize CSF amino acid profiles (Ferraro and Hare 1985; Hare and Manyam 1980). This method, however, has a lower resolution than our technique. As a consequence the fluorimetric detection and the technical modifications of our method we can measure 36 compounds in one run with a tenfold increase of sensitivity compared to the conventional ninhydrin method (Gerrits et al 1989).

Concerning the reference values of purines and pyrimidines in CSF it can be stated that few data are available from literature. Abnormal concentrations of some purines and pyrimidines have been reported in a few cases of disturbances in purine and pyrimidine metabolism (Lesch and Nyhan 1964; Jaeken and Van den Berghe 1984;

Bakkeren et al 1984). Also some reports have been published on the oxypurines in CSF in relation to hypoxia and cerebral ischaemia (Manzke et al 1981; Harkness and Lund 1983).

In our patients with arginosuccinic aciduria we demonstrated aberrations in pyrimidines limited to CSF (Gerrits et al 1993). We were particularly interested in the CSF concentrations of pyrimidines, because the urea cycle is present in brain (Sporn et al 1959). Therefore, CSF could be a sensitive indicator fluid in which to detect urea cycle defects distal to carbamyl-phosphate synthetase (E.C. 2.7.2.2.) by elevated pyrimidine concentrations. In arginosuccinic aciduria we demonstrated, elevated concentrations of uridine and pseudouridine limited to CSF. Pseudouridine, a tRNA catabolyte and metabolic end product, is associated with tissue destruction.

Our results in patients with chronic renal failure are striking (Gerrits et al 1991). Patients with uraemia are prone to develop a uraemic encephalopathy. Our results show that there are metabolic aberrations in uraemia limited to the cerebral compartment. The concentrations of cytidine in CSF are in our patients even greater than the plasma concentration. This could possible give some clues for the still not understood uraemic encephalopathy. Cytidine is involved in membrane metabolism as has been pointed out (Gerrits et al 1991). We demonstrated that CAPD had no influence on these elevated concentrations.

An impaired transport of amino acids can explain the lowered concentration of most amino acids in CSF (Cangiano et al 1988).

The screening for metabolic disorders in patients with unexplained mental retardation revealed only 2 patients with reproducible abnormalities in CSF amino acids. These patients showed a decrease of homocarnosine and two unknown peaks limited to CSF. The bulk of patients with retardation showed no reproducible abnormalities in either amino acids, or purines and pyrimidines. Probably, when we are able to describe different clinical subgroups in mental retardation, it will also be possible to discern statistically significant abnormalities, e.g. as we are able to describe abnormalities in amino acids in uraemia.

In our view CSF can be used to study metabolic diseases suspected to be limited to the cerebral compartment. For screening of mental retardation an additional value is only present for a small number of patients. In view of the large number of retarded children studied the results are rather poor. This is of particular importance since in mental retardation an increasing number of clinicians are requiring metabolic screening of CSF.

In this article we have given an overview of our results of using CSF amino acid and purine and pyrimidine contents as tools in the study of metabolic brain diseases. In mental retardation, clinical and biochemical findings should point to a specific cerebral metabolic involvement before CSF investigation of amino acids, purines and pyrmidines should be performed.

REFERENCES

Applegarth DA, Edelsten AD, Wong LTK, Morrison BJ (1979) Observed range of assay values for plasma and cerebrospinal fluid amino acid levels in infants and children aged 3 months to 10 years. *Clin Biochem* **12**: 173–178.

Bakkeren JAJM, De Abreu RA, Sengers RCA, Gabreëls FJM, Maas JM, Renier WO (1984) Elevated urine, blood and cerebrospinal fluid levels of uracil and thymine in a child with dehydrothymine dehydrogenase deficiency. *Clin Chim Acta* 14: 247–256.

Cangiano C, Cardelli-Cangiano P, Cascino A, Ceci F, Fiori A, Mulieri M, Muscaritoli M, Barberini C, Strom R, Fanelli FR (1988) Uptake of amino acids by brain microvessels from rats with experimental chronic renal failure. *J Neurochem* 51: 1675–1681.

Ferraro TN, Hare TA (1985) Free and conjugated amino acids in human CSF: influence of age and sex. *Brain Res* 338: 53–60.

Gerrits GPJM, Haagen AAM, De Abreu RA, Monnens LAH, Gabreëls FJM, Trijbels JMF, Theeuwes ALM, Van Baal JM (1988) Reference values for nucleosides and nucleobases in cerebrospinal fluid of children. *Clin Chem* 34: 1439–1442.

Gerrits GPJM, Trijbels JMF, Monnens LAH, Gabreëls FJM, De Abreu RA, Theeuwes AGM, Van Raay-Selten B. (1989) Reference values for amino acids in cerebrospinal fluid of children determined with ion-exchange chromatography using fluorimetric detection. *Clin Chim Acta* 182: 271–280.

Gerrits GPJM, Monnens LAH, De Abreu RA, Schröder CH, Trijbels JMF, Gabreëls FJM (1991) Disturbances of cerebral purine and pyrimidine metabolism in young children with chronic renal failure. *Nephron* 58: 310–314.

Gerrits GPJM, Gabreëls FJM, Monnens LAH, De Abreu RA, van Raaij-Selten B, Niezen-Koning KE, Trijbels JMF (1993). Argininosuccinic aciduria: clinical and biochemical findings in three children with the late onset form, with special emphasis on cerebrospinal fluid findings of amino acids and pyrmidines. *Neuropediatrics,* 24: 15–18.

Hare TA, Manyam NVB (1980) Rapid and sensitive ion-exchange fluorimetric measurement of gamma-aminobutyric acid in physiological fluids. *Anal Biochem* 101: 349–355.

Harkness RA, Lund RJ (1983) Cerebrospinal fluid concentrations of hypoxanthine, uridine and inosine: high concentrations of the ATP metabolite, hypoxanthine after hypoxia. *J Clin Pathol* 36: 1–8.

Heiblim DI, Evans HE, Glass L, Agbayani MM (1978) Amino acid concentrations in cerebrospinal fluid. *Arch Neurol* 35: 765–768.

Honda T (1984) Amino acid metabolism in the brain with convulsive disorders. Part 3: free amino acid patterns in cerebrospinal fluid in infants and children with convulsive disorders. *Brain Dev* 6: 27–32.

Jaeken J, Van Den Berghe G (1984) An infantile autistic syndrome characterised by the presence of succinylpyrines in body fluids. *Lancet* 2: 1058–1061.

Jaeken J, Casaer P, Haegele KD, Schechter PJ (1990) Review: Normal and abnormal central nervous system GABA metabolism in childhood. *J Inher Metab Dis* 13: 793–801.

Lesch M, Nyhan WLK (1964) A familial disorder of uric acid metabolism and central nervous system function. *Am J Med* 36: 561–570.

Manzke H, Staemmler W, Dörner K (1981) Increased nucleotide catabolism after cerebral convulsions. *Neuropediatrics* 12: 119–131.

Nyhan WL (1989) Nonketotic hyperglycinemia. In Scriver CR, Beaudet AL, Sly WS, Valle D (eds) *The metabolic basis of inherited disease.* McGraw-Hill, New York, 1989, pp 743–753.

Scriver RC, Rosenberg LE (1973) *Amino acid metabolism and its disorders.* Saunders, Philadelphia, p. 245.

Sporn MB, Dingman W, Defalco A, Davies RK (1959) The synthesis of urea in the living rat brain. *J Neurochem* 5: 62–67.

Van Erven PMM, Gabreëls FJM, Ruitenbeek W, Renier WO, Lamers KJB, Slooff JL (1987) Familial Leigh's syndrome. Association with a defect in oxidative metabolism probably restricted to brain. *J Neurol* 234: 215–219.

J. Inher. Metab. Dis. 16 (1993) 676–690
© SSIEM and Kluwer Academic Publishers. Printed in the Netherlands

Abnormalities of Biogenic Amine Metabolism

K. HYLAND

Metabolic Disease Center, Baylor Research Institute, 3812 Elm Street, PO Box 710699, Dallas, TX 75226, USA

Summary: The term biogenic amine is an umbrella term that encompasses all amines with an origin in biological processes. This review will be restricted to the biogenic amine abnormalities that affect the metabolism of serotonin and the catecholamines. The synthesis and catabolism of these neurotransmitters are outlined, and a summary is given of the neurological details, biochemical features, and treatment of the inborn errors that primarily affect their metabolism. An idea is also developed that proposes that abnormalities of biogenic amine metabolism are far more common than is currently considered, and that the search for these problems may be appropriate in any neonate or infant who presents with neurological problems of unknown origin.

The neurological symptoms that develop in infancy due to a combined deficiency of serotonin and the catecholamines (monoamines) are fairly characteristic and have been established, for the most part, from the study of inborn errors that affect tetrahydrobiopterin (BH_4) metabolism. The combined neurotransmitter defect arises as BH_4 is the cofactor for both tyrosine hydroxylase (EC 1.14.16.2) and tryptophan hydroxylase (EC 1.14.16.4) (Figure 1), these being the rate-limiting steps in the synthesis of dopamine and serotonin. The very early symptoms of BH_4 (and hence monoamine) deficiency are, however, non-specific, and diagnosis would probably not be made were it not for the fact that BH_4 is also the cofactor for phenylalanine hydroxylase (EC 1.14.16.1) (Figure 1). Absence of BH_4 therefore results in phenylketonuria (PKU) which is, for the most part, detected during neonatal screening. If PKU is not present, the possibility of a neurotransmitter deficiency in a child with non-specific neurological problems is usually not considered. The detection of aromatic L-amino acid decarboxylase (EC 4.1.1.28) deficiency (Hyland and Clayton 1990; Hyland et al 1992), a disease that also causes defective synthesis of serotonin and the catecholamines (Figure 1), has demonstrated that primary defects in serotonin and dopamine metabolism can occur in the absence of PKU and that an alternative approach is required in order to detect such problems. Unfortunately, there is a lack of early clinical markers in these cases. It is hoped that this review, by assembling the information on the current known inborn errors of serotonin and catecholamine metabolism and by discussing the possibility of other defects, will lead to a systematic investigation of monoamine metabolism in all neonates and infants with non-specific neurological disease of unknown origin. In this manner the true incidence of these

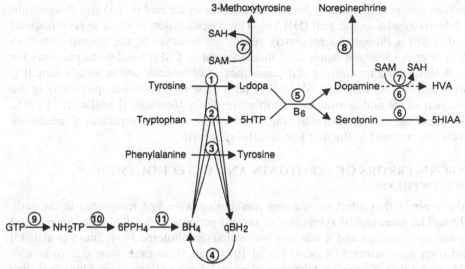

Figure 1 Synthesis and catabolism of serotonin and dopamine. 1 = tyrosine hydroxylase; 2 = tryptophan hydroxylase; 3 = phenylalanine hydroxylase; 4 = dihydropteridine reductase; 5 = aromatic L-amino-acid decarboxylase; 6 = monoamine oxidase; 7 = catechol O-methyltransferase; 8 = dopamine β-hydroxylase; 9 = GTP cyclohydrolase; 10 = 6-pyruvoyltetrahydropterin synthetase; 11 = sepiapterin reductase; 5HTP = 5-hydroxytryptophan; B_6 = pyridoxal-5'-phosphate; HVA = homovanillic acid; 5HIAA = 5-hydroxyindolacetic acid; BH_4 = tetrahydrobiopterin; qBH_2 = quinoid dihydrobiopterin; SAM = S-adenosylmethionine; SAH = S-adenosylhomocysteine

abnormalities can be established and a clearer picture of the early clinical features can be ascertained. It is important that this be achieved, as all the data to date suggest that the neurological features associated with these conditions can be ameliorated provided that they are detected early.

SYNTHESIS AND CATABOLISM OF SEROTONIN AND DOPAMINE

The pathways for the metabolism of serotonin and dopamine are shown in Figure 1. The neurotransmitters are formed from tryptophan and tyrosine in reactions catalysed by tryptophan hydroxylase and tyrosine hydroxylase. These two enzymes are rate limiting for the synthesis of their respective neurotransmitters (Kaufman 1981) and both require molecular oxygen and BH_4 for their activity. Following hydroxylation of the tyrosine and tryptophan, the L-dopa and 5-hydroxytryptophan products are decarboxylated by the pyridoxine-dependent aromatic L-amino acid decarboxylase to form the active neurotransmitters. In specific groups of cells, dopamine is further hydroxylated by the action of dopamine β-hydroxylase (EC 1.14.17.1) to form noradrenaline, which can then be methylated to yield adrenaline.

The main routes of catabolism of serotonin and the catecholamines involve either methylation of the catechol moiety via the action of catechol O-methyltransferase (EC 2.1.1.6) or the formation of acidic metabolites by aldehyde dehydrogenase (EC 1.2.1.3) and monoamine oxidase (EC 1.4.3.4) (Figure 1). In man, the major central nervous system metabolite of dopamine is homovanillic acid (HVA) and of serotonin is 5-hydroxyindoleacetic acid (5HIAA). The concentration of these in cerebrospinal fluid (CSF) is thought to accurately reflect the turnover of the neurotransmitters (Wester et al 1990) and hence their measurement in CSF provides the mainstay for the detection and monitoring of diseases that affect biogenic amine metabolism. It is important to note that the CSF level of both these metabolites drops rapidly in the first year of life and continues to drop more slowly thereafter (Langlais et al 1985). The use of age-related reference ranges is therefore essential, especially if metabolite levels are analysed in the first few months after birth.

INBORN ERRORS OF SEROTONIN AND CATECHOLAMINE BIOSYNTHESIS

Inborn errors that affect monoamine metabolism were first recognized in the early 1970s. The neurological symptoms of several patients with PKU failed to respond to dietary treatment and it was first postulated (Bartholomé 1974; Smith et al 1975) and then demonstrated (Kaufman et al 1975) that these cases were due to lack of BH_4 and not to a primary deficiency of phenylalanine hydroxylase. Since that time over 200 cases have been reported worldwide (Dhondt 1991), the metabolic defects being shown to exist in both the biosynthetic and recycling pathways of BH_4 metabolism. Other reported inborn errors that affect monoamine metabolism include AADC deficiency (Hyland and Clayton 1990), dopamine β-hydroxylase deficiency (Man in't Veld et al 1987), and monoamine oxidase deficiency (in association with Norrie disease) (Collins et al 1992).

Synthesis of tetrahydrobiopterin: Detailed accounts of the biosynthesis and metabolism of BH_4 have been previously reviewed (Nichol et al 1985; Duch et al 1991). BH_4 is synthesized from GTP in a multi-step pathway involving neopterin, and tetrahydropterin intermediates (Figure 1). GTP is converted to dihydroneopterin triphosphate by GTP cyclohydrolase (EC 3.5.4.16). The action of 6-pyruvoyltetrahydropterin synthetase leads to the loss of the phosphates and an internal oxidoreduction to form 6-pyruvoyltetrahydropterin. The two keto groups on the side-chain are then reduced to form BH_4. Whether these two reductions are both performed by sepiapterin reductase (EC 1.1.1.153) or whether another enzyme, 6-pyruvoyltetrahydropterin reductase is also required is still unclear (Curtius et al 1986).

During the hydroxylation of tyrosine and tryptophan, BH_4 is oxidized to quinonoid dihydrobiopterin and this is reduced back to the active form by dihydropteridine reductase (DHPR) (EC 1.6.99.7).

Defects of tetrahydrobiopterin metabolism

The inborn errors affecting BH_4 metabolism have been reviewed previously (Scriver et al 1990; Smith 1990) and a very comprehensive account of most of the detected cases is provided in the International Register of Tetrahydrobiopterin Deficiencies, which is readily available on request (Dhondt 1991).

Today, most of the new cases with defects of BH_4 metabolism are detected in the neonatal period. They are initially detected because of a finding of PKU, follow-up studies (see below) being necessary to show the presence of the pterin defect. Occasionally patients still present later; normally they originate either from countries where screening for PKU is absent or where further testing for pterin defects is not performed.

There are three reported enzymopathies that affect BH_4 metabolism and the turnover of serotonin and dopamine. GTP cyclohydrolase and 6-pyruvoyltetrahydropterin synthetase deficiency affect the biosynthetic pathway, and DHPR deficiency prevents recycling of the cofactor. In addition, a presumed deficiency of phenylalanine hydroxylase stimulator protein has also been described (Dhondt et al 1988; Curtius et al 1990). The lack of this enzyme leads to the formation of 7-biopterins (primapterin) (as opposed to the normal 6-biopterins). These inhibit phenylalanine hydroxylase and therefore lead to hyperphenylalaninaemia (Curtius et al 1990); however, the CSF levels of HVA and 5HIAA are normal and neurological symptoms do not seem to be present after correction of the raised blood phenylalanine (Dhondt et al 1988; Blaskovics and Giudici 1988).

Expanding on the idea developed in the International Register of BH_4 defects (Dhondt 1991), all defects of BH_4 metabolism can be simply divided into typical and atypical forms, the definition being dependent on whether or not treatment of a biogenic amine deficiency is necessary.

Diagnosis: The possibility of a defect in BH_4 metabolism should be considered in all cases where hyperphenylalaninaemia has been found on neonatal screening and also if characteristic neurological signs are present in a child who presents later. The methods for diagnosis of defects in BH_4 metabolism have been extensively reviewed previously (Hyland and Howells 1988; Niederwieser et al 1982a; Smith 1990). They rely on the appearance of characteristic pterin profiles in urine from children affected with the different defects (Figure 2). Urine is oxidized to convert BH_4, quinonoid dihydropterin and BH_2 to biopterin (total biopterin) and dihydroneopterin to neopterin (total neopterin), and then the biopterin and neopterin are separated by HPLC. In all cases, concentrations of biopterin and neopterin are greatly reduced in GTP cyclohydrolase deficiency. Patients with 6-pyruvoyltetrahydropterin synthetase deficiency have decreased biopterin and elevated neopterin, and those with DHPR deficiency have elevated biopterins with normal neopterins. The elevation of biopterin in DHPR deficiency is due to accumulation of $7,8\text{-}BH_2$ that is formed following tautomerization of quinonoid dihydropterin. 7-Biopterin is excreted in equal amounts to the natural 6-biopterin isomer in patients with a putative deficiency of phenylalanine stimulator protein (Dhondt et al 1988).

A BH_4 loading test can be used to confirm the diagnosis of GTP cyclohydrolase, 6-pyruvoyltetrahydropterin synthetase deficiency or phenylalanine hydroxylase stimulator protein deficiency. Oral administration of 2–20 mg/kg of BH_4 has led to a drop in plasma phenylalanine concentration in all cases (Dhondt 1987, 1991). It is obviously important that this test be performed before the commencement of a low-phenylalanine diet.

Several problems can arise using these diagnostic techniques. Hyperphenylalanin-

HYPERPHENYLALANINEMIA

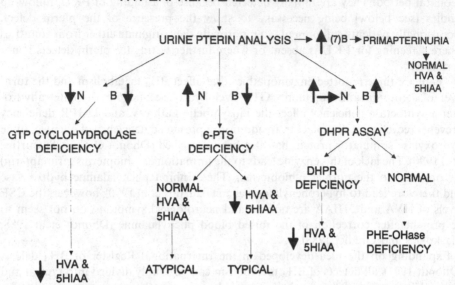

Figure 2 Urine neopterin, biopterin and CSF neurotransmitter metabolite pattern in inborn errors affecting tetrahydrobiopterin metabolism. N = total neopterin; B = total biopterin; HVA = homovanillic acid; 5HIAA = 5-hydroxyindoleacetic acid; DHPR = dihydropteridine reductase; 6-PTS = 6-pyruvoyltetrahydropterin synthetase; Phe-OHase = phenylalanine hydroxylase

aemia, no matter what the cause, leads to an elevation of plasma and urine total biopterin (Dhondt et al 1981) and it is sometimes difficult to distinguish between phenylalanine hydroxylase deficiency and DHPR deficiency (Kaufman 1986). A BH$_4$ loading test is also not reliable in DHPR deficiency where some cases fail to respond even with a 20 mg/kg dose of BH$_4$ (Endres et al 1987). It is therefore necessary to measure DHPR activity in all cases of PKU. This can be reliably accomplished using a dried blood spot (Arai et al 1982). Urine analysis of pterins is also unable to distinguish between typical and atypical 6-pyruvoyltetrahydropterin synthetase deficiency. The atypical forms have PKU, demonstrating that BH$_4$ metabolism is abnormal peripherally; however, concentrations of 5HIAA and HVA are normal within CSF and neurological symptoms are absent following correction of the hyperphenylalaninaemia. Measurement of CSF HVA and 5HIAA is therefore essential for differentiation between the two forms. In the cases where it has been measured, the levels of total biopterin in CSF have been within the normal range, whereas there is still an elevation of neopterin, thus demonstrating the existence of the abnormality in BH$_4$ metabolism (Dhondt 1991).

Typical defects — neurological signs: The neurological symptoms of untreated 'typical' cases of GTP cyclohydrolase, 6-pyruvoyltetrahydropterin synthetase and DHPR deficiency are virtually identical. Patients present between 2 and 8 months of age with a fairly well-characterized neurological syndrome. Symptoms include hypersalivation and temperature disturbance (in the absence of infection), pinpoint pupils, oculogyric crises, hypokinesis, distal chorea, truncal hypotonia, swallowing difficulties, drowsiness, and irritability. In addition there may be microcephaly, progressive neurological deterioration, developmental delay, and convulsions (grand mal or myoclonic) (Smith 1990; Dhondt 1991). Abnormal signs in the neonatal period can include poor sucking, decreased spontaneous movements and microcephaly, particularly in 6-pyruvoyltetrahydropterin synthetase deficiency. The higher incidence of low birth weight (Smith and Dhondt 1985) in 6-pyruvoyltetrahydropterin synthetase deficiency suggests that *in utero* damage may occur. The presence of high activities of dihydrofolate reductase and 5,10-methylenetetrahydrofolate reductase (which provide alternate pathways for the regeneration of BH_4) has been put forward as the reason why such damage is less common in DHPR deficiency (Kaufman 1986).

In addition to the above symptoms, some patients with DHPR deficiency develop long-tract signs associated with multifocal, perivascular demyelination in the subcortical white matter, together with perivascular calcification that is located mainly in the basal ganglia but is also present in areas of white and grey matter (Smith et al 1985). These changes are not seen in BH_4 synthesis defects and they are thought to arise following the development of a central nervous system folate deficiency. There is speculation that the folate deficiency occurs as a result of inhibition of folate metabolism by the dihydropterins that accumulate in the disease (Smith et al 1985).

Treatment of typical defects: There have been a few reports of successful treatment of the neurological symptoms of 6-pyruvoyltetrahydropterin synthetase deficiency with BH_4 monotherapy (Niederwieser et al 1982b; Kaufman et al 1983) but in general central monoamine metabolism has been unaffected. We have shown in the *hph*-1 mouse mutant (which is deficient in GTP cyclohydrolase, and has half normal brain BH_4 concentrations) that the level of tyrosine hydroxylase within the brain is similarly reduced (Hyland and Engle, unpublished). These data suggest that BH_4 may regulate the steady-state concentration of tyrosine hydroxylase. If this is the case, it is not surprising that limited short-term trials of BH_4 monotherapy have been ineffective.

Precursor therapy with L-dopa and 5-hydroxytryptophan, in conjunction with a peripheral decarboxylase inhibitor, is the main treatment for the biogenic amine deficiency. As phenylalanine competes with L-dopa and 5-hydroxytryptophan for passage across the blood–brain barrier, maintenance of a consistent, low plasma phenylalanine level is vital if large fluctuations in response to the replacement therapy are to be avoided. Normal plasma phenylalanine concentrations in 6-pyruvoyltetrahydropterin synthetase deficiency and GTP cyclohydrolase deficiency can be maintained by administration of small doses of BH_4 (2–4 mg/kg per day), although the dose required is variable. This therapy is preferable to the low-phenylalanine diet. Unfortunately, the use of BH_4 does not seem feasible for the treatment of the hyperphenylalaninaemia in DHPR deficiency as large doses are required in the absence of the recycling system (Kaufman 1986). Some patients do respond, however (Smith et al 1985). Also, theoretically, administration of BH_4 may exacerbate the folate deficiency that develops if it eventually adds to the already high levels of BH_2.

The amount of L-dopa and 5-hydroxytryptophan required for effective correction of the amine deficiency is extremely variable and must be determined in each patient on the basis of both the clinical and neurochemical response. It is therefore necessary to regularly monitor the CSF levels of HVA and 5HIAA, especially as many of the clinical symptoms of amine deficiency are similar to those of an amine excess.

The folate abnormality that appears in DHPR deficiency develops in an insidious manner. Because of the devastating symptoms that accompany its appearance, treatment with folinic acid is recommended from the time of diagnosis, even in the absence of any measurable abnormality in folate metabolism. Folic acid should not be used as it has been shown to exacerbate the neurological symptoms, at least in a child who was already folate deficient (Smith et al 1985).

Atypical defects

Atypical defects affecting 6-pyruvoyltetrahydropterin synthetase (Hoganson et al 1984) and the phenylalanine hydroxylase-stimulating protein (primapterinuria) (Curtius et al 1990) have been described. These defects are again detected during neonatal screening for PKU. There are usually none of the prominent signs of anormal monoamine metabolism and the concentrations of HVA and 5HIAA in CSF are essentially normal. In most cases, the only treatment required is a correction of the hyperphenylalaninaemia with either a low-phenylalanine diet or BH_4; however, there has been a report of an atypical form of 6-pyruvoyltetrahydropterin synthetase deficiency progressing to give a central phenotype and it was suggested that all patients should be re-evaluated in terms of their central amine status later in infancy (Ponzone et al 1990). As pointed out by Dhondt, there is obvious heterogeneity within this atypical group, and information of the long-term prognosis is unavailable; it therefore seems prudent to treat all cases with low-dose BH_4 therapy and to carefully monitor developmental progress.

Aromatic L-amino acid decarboxylase deficiency

There has been a single diagnosis of male homozygous twins with aromatic L-amino acid decarboxylase deficiency (Hyland and Clayton 1990; Hyland et al 1992). Given the post-natal survival of these twins it seems likely that the lack of detection of other cases may be due to non-recognition of a biogenic amine deficiency in the absence of hyperphenylalaninaemia and an accompanying pterin defect.

Neurological symptoms: The twins presented around 2 months of age with generalized hypotonia, developmental delay and paroxysmal movements with oculogyric crises. Anticonvulsant therapy was ineffective. At 9 months there was central and peripheral hypotonia, unstable temperature, fine chorea of distal limbs, and continuing oculogyric crises. Head circumference, length and weight were within the normal range and electroencephalography was normal even during paroxysmal movements. Brain imaging (CAT and MRI) showed only cerebral atrophy.

There is probably a fine line between clinically significant and non-significant abnormalities of L-amino acid decarboxylase. In the rat the enzyme is not rate-limiting for the synthesis of either dopamine or serotonin in the CNS. This also appears to be the case in the human as the parents of the affected index cases had plasma aromatic L-amino acid decarboxylase activities of 16% and 19% of control values, and yet they had a totally normal phenotype (Hyland et al 1992).

Diagnosis: It is unlikely that the diagnosis of aromatic L-amino acid decarboxylase deficiency would have been made using standard methods for the investigation of neurometabolic disease. The disease was only detected because a protocol had already been established that called for CSF to be obtained and monoamine metabolism to be investigated in cases of unexplained neurological disease. Analysis of CSF demonstrated very low concentrations of CSF HVA and 5HIAA in the absence of an abnormality of BH_4 or phenylalanine metabolism (Table 1). Further studies showed that the defect was not localized to the central nervous system, as concentrations of whole blood serotonin and of the plasma catecholamines were also very low. The absence of a pterin defect and the global nature of the disease suggested a deficiency of aromatic L-amino acid decarboxylase, as this is the only other common feature of the synthesis of both serotonin and the catecholamines (Figure 1). Very low activity of L-dopa decarboxylase in plasma and of 5-hydroxytryptophan decarboxylase and L-dopa decarboxylase in a liver biopsy sample confirmed the diagnosis (Hyland and Clayton 1990; Hyland et al 1992).

Deficiency of aromatic L-amino acid decarboxylase leads to accumulation of L-dopa, 5HTP and 3-methoxytyrosine in CSF, plasma and urine. The latter compound accumulates following methylation of L-dopa (Sharpless et al 1971) and it is likely that the elevation of 3-methoxytyrosine will provide the diagnostic marker for all future cases of the disease. Like HVA and 5HIAA, the concentrations of 3-methoxytyrosine in CSF and plasma decrease rapidly in the first year of life (Hyland and Clayton 1992) and values must be related to age-matched reference ranges. The plasma levels of 3-methoxytyrosine in the index cases were around 10 μmol/L, which is about the detection limit of many amino acid analysers, and unless it is specifically looked for, it is likely to be missed. The major metabolite of 3-methoxytyrosine is vanillactic acid. A 20-fold elevation of vanillactic acid was seen using standard

Table 1 Cerebrospinal fluid metabolite concentrations in the index cases of aromatic L-amino acid decarboxylase deficiency

Metabolite	Twin 1	Twin 2	Normal range	Units
Homovanillic acid	60	60	240–851	nmol/L
5-Hydroxyindoleacetic acid	10	21	168–451	nmol/L
Tetrahydrobiopterin	31	26	19–56	nmol/L
Neopterin	13.9	9.9	7–65	nmol/L
Phenylalanine	6.3	7.2	5–14.2	μmol/L
L-dopa	305	311	< 25	nmol/L
3-Methoxytyrosine	1650	1585	< 50	nmol/L
5-Hydroxytryptophan	139	139	< 10	nmol/L

GC–MS analysis of urinary organic acids. Aromatic L-amino acid decarboxylase deficiency should therefore be considered if an inappropriate rise in urinary vanillactic acid is detected.

HVA, 5HIAA and 3-methoxytyrosine can now be measured simultaneously in CSF (Hyland and Clayton 1992) using HPLC with electrochemical detection. An extremely characteristic profile is found in L-amino acid decarboxylase deficiency, with the appearance of very low levels of 5HIAA and HVA and a large elevation in 3-methoxytyrosine (Figure 3). Analysis of CSF using this analytical technique should allow detection of L-amino acid decarboxylase deficiency in the future and allow the disease to be distinguished from other problems affecting monoamine metabolism.

Treatment of L-amino acid decarboxylase deficiency: The index cases of L-amino acid decarboxylase deficiency were treated with a combination of pyridoxine (cofactor for L-amino acid decarboxylase; 100 mg b.d.), bromocriptine (dopamine agonist; 2.5 mg b.d.) and tranylcypromine (monoamine oxidase inhibitor; 4 mg b.d.). Pyridoxine (50 mg b.d.) alone had no clinical effect, but did lead to a drop in CSF 3-methoxytyrosine and L-dopa, and to an increase in HVA levels. An increase of the dose

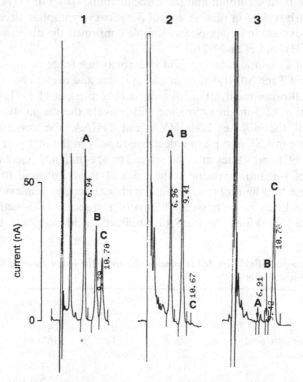

Figure 3 Electrochemical detection of 5HIAA, HVA and 3-methoxytyrosine in CSF. 1 = standard, 500 nmol/L 5HIAA, HVA and 3-methoxytyrosine. 2 = normal CSF. 3 = CSF from patient with aromatic L-amino acid decarboxylase deficiency. A = 5HIAA. B = HVA. C = 3-methoxytyrosine. Injection volume = 10 μl

to 100 mg b.d. caused a further decrease in the abnormal metabolite concentrations. Bromocriptine alone stopped the oculogyric crises but had little effect on spontaneous movement. Tranylcypromine improved muscle tone, increased spontaneous movement, and reduced sweating. Before treatment, growth had stopped. One year of treatment led to a return of the twins' weight and length to the 10th centile and their head circumferences to the 50th centile. The failure of pyridoxine to affect the clinical pattern should not prevent trials of monotherapy with this cofactor in any future cases, as binding defects for the cofactor may exist similar to those seen in some cases of cystathionine β-synthetase deficiency.

Prenatal diagnosis: Prenatal diagnosis of aromatic L-amino acid decarboxylase deficiency is possible; unfortunately it currently requires the analysis of aromatic L-amino acid decarboxylase activity in fetal liver, as neither cultured amniocytes nor chorionic villi show any detectable enzyme activity (Hyland et al 1992). The single prenatal diagnosis performed to date showed normal aromatic L-amino acid decarboxylase activity and no accumulation of 3-methoxytyrosine in amniotic fluid (Hyland et al 1992). A normal female child was delivered at term and plasma activity was within the normal range. Although not demonstrated in this case, it is probable that an affected fetus would excrete 3-methoxytyrosine into the amniotic fluid and that prenatal diagnosis may be possible by measurement of this metabolite, without the need for the liver biopsy.

Dopamine β-hydroxylase deficiency: Dopamine β-hydroxylase deficiency has only been diagnosed in the adult period when patients present with orthostatic hypotension (Robertson et al 1986; Man in't Veld et al 1987; Biaggioni et al 1990). Close examination of the history suggests, however, that symptoms are manifest even in the neonatal period. The most common finding is congenital bilateral ptosis; other features have included hypoglycaemia (possibly following adrenomedullary failure), slight skeletal hypotonia and fainting in early infancy. Two of the reported cases were treated, inappropriately, with anticonvulsants (Robertson et al 1986; Biaggioni et al 1990).

It is interesting to speculate whether these reports represent the 'worst case scenario' or whether they are mild forms of the disease. More major defects may have passed undiagnosed because the neurological symptoms that might be associated with the deficiency in infancy or the neonatal period are to date unknown. If these cases do exist, the analysis of amine metabolites in CSF would again be likely to establish the diagnosis. In the late-onset cases, dopamine β-hydroxylase deficiency leads to a build up of dopamine, and hence an elevation of HVA and 3-methoxytyrosine.

UNDETECTED ABNORMALITIES OF SEROTONIN AND CATECHOLAMINE METABOLISM

The data describing the neurological characteristics of monoamine deficiency in children have been derived from patients with defects of BH_4 metabolism or with L-amino acid decarboxylase deficiency. In these diseases there is a balanced deficiency of both serotonin and the catecholamines. The neurological consequences of an isolated deficiency of either serotonin or the catecholamines, caused by a primary inborn error of either tryptophan hydroxylase or tyrosine hydroxylase is currently unclear, as these diseases have yet to be described. There is, however, a linear relationship between HVA and 5HIAA in CSF (Figure 4), which suggests that the metabolism of serotonin and dopamine is normally carefully integrated and that disruption of either one may have a far more devastating effect than a balanced loss of both. As such, it is possible that a total lack of either tyrosine hydroxylase or tryptophan hydroxylase may not be compatible with life; alternatively, partial or total deficiencies may occur, but again these have not been detected because of the lack of investigation of appropriate patient groups.

We have begun to investigate neonates and infants with seizure disorders to establish whether an imbalance in serotonin or catecholamine metabolism may play a role in the pathogenic mechanisms underlying these conditions. There are several old inconclusive reports describing the investigation of monoamine metabolites in CSF from children with seizures (Ito et al 1980; Shaywitz et al 1975; Silverstein and Johnston 1984) and a more recent report of an L-dopa-responsive seizure disorder (Sugie et al 1989). This case was only discovered because monoamine metabolites were measured in CSF. L-Dopa prevented the seizures and also led to a marked improvement in EEG. We have also investigated a patient with neonatal seizures who had every low HVA (143 nmol/L) and normal 5HIAA who responded to L-dopa therapy; unfortunately the patient was lost to follow-up. It is possible that these two cases had a defect in tyrosine hydroxylase. Proving such a hypothesis is likely to require the use of molecular techniques, as there are four types of mRNA for tyrosine hydroxylase, these being produced by alternative splicing from a single gene (Nagatsu et al 1991).

The results of the preliminary studies are shown in Figure 4. Seven of 14 cases investigated so far had HVA and 5HIAA concentrations that deviated from reference values, with the concentrations of the metabolites either being balanced but grossly elevated, or unbalanced with the level of one metabolite being inappropriate for the associated normal level of the other. These early investigations of monoamine metabolism in children with seizure disorders of unknown origin suggest that further studies are warranted.

SECONDARY ABNORMALITIES OF BIOGENIC AMINE METABOLISM

It is important to recognize that abnormalities of serotonin and catecholamine metabolism also occur secondarily to problems in other areas of metabolism. Changes in CSF amine metabolite levels have been reported in liver disease, viral infection (O'Kusky et al 1991), abnormalities of folate metabolism (Clayton et al 1986) Lesch–Nyhan syndrome (Lloyd et al 1981), urea-cycle disorders (Hyman et al 1987), Rett syndrome (Wenk et al 1991), and in deficiencies of arginase (Hyland et al 1985) and phenylalanine hydroxylase (Smith 1990). This list is not comprehensive, but does serve to point out that other possibilities should be considered before making the diagnosis of a primary abnormality of monoamine metabolism.

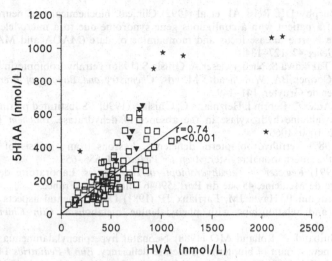

Figure 4 The relationship between homovanillic acid and 5-hydroxyindole acetic acid in CSF in controls and patients with seizure disorders. □ = controls; ★ = abnormal values (seizures); ▼ = normal values (seizures)

CONCLUSIONS

For several years our laboratory has been developing tests that allow the detection and localization of defects of serotonin, catecholamine and 1-carbon metabolism using the analysis of CSF as the primary investigational matrix. The range of analyses includes measurement of the precursor aromatic amino acids (Hyland et al 1986), monoamine metabolites (Hyland et al 1986; 1992), all the different oxidation states of the biopterins and neopterins (Howells and Hyland 1987), 5-methyltetrahydrofolate (Hyland and Surtees 1992), S-adenosylmethionine (Surtees and Hyland 1989) and total homocysteine (Hyland and Bottiglieri 1992). This series of biochemical tests can be used to accurately pinpoint all of the known defects of monoamine metabolism and will probably lead to the diagnosis of other as yet undescribed problems. For this to occur first requires that the clinician consider an abnormality of monoamine metabolism in all neonates and infants with neurological disease of unknown origin and that CSF be collected for analysis.

REFERENCES

Arai N, Narisawa K, Hayakawa H, Tada K (1982) Hyperphenylalaninemia due to dihydropteridine reductase deficiency: diagnosis by enzyme assays on dried blood spots. *Pediatrics* **70**: 426–430.
Bartholomé K (1974) A new molecular defect in phenylketonuria. *Lancet* **2**: 1580.
Biaggioni I, Goldstein DS, Atkinson T, Robertson D (1990) Dopamine beta-hydroxylase deficiency in humans. *Neurology* **40**: 370–373.
Blaskovics M, Giudici TA (1988) A new variant of biopterin deficiency. *N Engl J Med* **319**: 1611–1612.
Clayton PT, Smith I, Harding B, Hyland K, Leonard JV, Leeming RJ (1986) Subacute combined degeneration of the cord, dementia and parkinsonism due to an inborn error of folate metabolism. *J Neurol Neurosurg Psychiatr* **49**: 920–927.

Collins FA, Murphy DL, Reiss AL et al (1992) Clinical, biochemical and neuropsychiatric evaluation of a patient with a contiguous gene syndrome due to a microdeletion Xp11.3 including the Norrie disease locus and monoamine oxidase (MAOA and MAOB) genes. *Am J Med Genet* **42**: 127–134.

Curtius H-Ch, Takikawa S, Niederwieser A, Ghisla S (1986) Tetrahydrobiopterin biosynthesis in man. In Cooper BA, Whitehead VM, eds. *Chemistry and Biology of Pteridines 1986.* Berlin: Wlater de Gruyter, 141–149.

Curtius H-Ch, Adler C, Rebrin I, Heizmann C, Ghisla S (1990) 7-Substituted pterins: formation during phenylalanine hydroxylase in the absence of dehydratase. *Biochem Biophys Res Commun* **172**: 1060–1066.

Dhondt JL (1987) Tetrahydrobiopterin deficiency. Lessons from analysis of 90 patients collected in the international register. *Arch Fr Pediatr* **44**: 655–659.

Dhondt JL (1991) *Register of Tetrahydrobiopterin Deficiencies.* Laboratoire de Biochimie, Faculte Libre de Medicine, 45 rue du Port, 59046 Lille Cedex, France.

Dhondt JL, Ardouin P, Hayte JM, Farriaux JP (1981) Developmental aspects of pteridine metabolism and relationships with phenylalanine metabolism. *Clin Chim Acta* **116**: 143–152.

Dhondt JL, Guibaud P, Rolland MO (1988) Neonatal hyperphenylalaninemia presumably caused by a new variant of biopterin synthetase deficiency. *Eur J Pediatrics* **147**: 153–157.

Duch DS, Smith GK (1991) Biosynthesis and function of tetrahydrobiopterin. *J Nutr Biochem* **2**: 411–423.

Endres W, Ibel H, Kierat L, Blau N, Curtius H-Ch (1987) Tetrahydrobiopterin and 'non-responsive' dihydropteridine reductase deficiency. *Lancet* **2**: 223.

Hoganson G, Berlow S, Kaufman S et al (1984) Biopterin synthesis defects: Problems in diagnosis. *Pediatrics* **74**: 1004–1011.

Howells DW, Hyland K (1987) Direct analysis of tetrahydrobiopterin in cerebrospinal fluid by high-performance liquid chromatography with redox electrochemistry: prevention of autoxidation during storage and analysis. *Clin Chim Acta* **167**: 23–30.

Hyland K, Bottiglieri T (1992) Measurement of total plasma and cerebrospinal fluid homocysteine by fluorescence following high performance liquid chromatography and precolumn derivatization with o-phthaldialdehyde. *J Chromatogr* **579**(1): 55–62.

Hyland K, Clayton PT (1990) Aromatic amino acid decarboxylase deficiency in twins. *J Inher Metab Dis* **13**: 301–304.

Hyland K, Clayton PT (1992) Aromatic L-amino acid decarboxylase deficiency: diagnostic methodology. *Clin Chem* **38**: 2405–2410.

Hyland K, Howells DW (1988) Analysis and clinical significance of pterins. *J Chromatogr* **429**: 95–121.

Hyland K, Surtees R (1992) Measurement of 5-methyltetrahydrofolate in cerebrospinal fluid using HPLC with coulometric electrochemical detection. *Pteridines* **3**: 149–150.

Hyland K, Smith I, Clayton PT, Leonard JV (1985) Impaired neurotransmitter amine metabolism in arginase deficiency. *J Neurol Neurosurg Psychiatr* **48**: 1188.

Hyland K, Howell DW, Smith I (1986) An isocratic high-performance liquid chromatographic system for the investigation of abnormalities of neurotransmitter amine, biopterin, and aromatic amino acid metabolism in cerebrospinal fluid using sequential coulometric electrochemical and fluorescence detection. In Joseph MH, Fillenz M, Macdonald IA, Marsden C, eds. *Monitoring Neurotransmitter Release During Behaviour.* Oxford, UK: A Ellis Howard, 233–238.

Hyland K, Surtees R, Rodeck C, Clayton PT (1992) Aromatic L-amino-acid decarboxylase deficiency: clinical features, diagnosis and treatment of a new inborn error of neurotransmitter amine synthesis. *Neurology* **42**: 1980–1988.

Hyman SL, Porter CA, Page TJ et al (1987) Behavioral management of feeding disturbances in urea cycle and organic acid disorders. *J Pediatr* **111**: 558–562.

Ito M, Okuno T, Mikawa H (1980) Elevated homovanillic acid in cerebrospinal fluid of children with infantile spasms. *Epilepsia* **21**: 387–392.

Kaufman S (1981) Regulatory properties of pterin dependent hydroxylases: variations on a theme. In Usdin E, Weiner N, Youdim MBH, eds. *Function and Regulation of Monoamine Enzymes.* New York: Macmillan, 165–173.

Kaufman S (1986) Unsolved problems in diagnosis and therapy of hyperphenylalaninemia caused by defects in tetrahydrobiopterin metabolism. *J Pediatr* 109: 572–578.

Kaufman S, Holtzman NA, Milstein S, Butler IJ, Krumholz A (1975) Phenylketonuria due to deficiency of dihydropteridine reductase. *N Engl J Med* 293: 785–790.

Kaufman S, Kapatos G, Rizzo WB, Schulman JD, Tamarkin L, Van Loon GR (1983) Tetrahydropterin therapy for hyperphenylalaninaemia caused by defective synthesis of tetrahydrobiopterin. *Ann Neurol* 14: 308–315.

Langlais PJ, Walsh FX, Bird ED, Levy HL (1985) Cerebral fluid neurotransmitter metabolites in neurologically normal infants and children. *Pediatrics* 75: 580–586.

Lloyd KG, Hornykiewicz O, Davidson L et al (1981) Biochemical evidence of dysfunction of brain neurotransmitters in the Lesch–Nyhan syndrome. *N Engl J Med* 305: 1106–1111.

Man in't Veld AJ, Boomsma F, Moleman P, Schalekamp MADH (1987) Congenital dopamine-beta-hydroxylase deficiency. *Lancet* 1: 183–188.

Nagatsu T (1991) Genes for human catecholamine-synthesizing enzymes. *Neurosci Res* 12: 315–345.

Nichol CA, Smith GK, Duch DS (1985) Biosynthesis and metabolism of tetrahydrobiopterin and molybdopterin. *Annu Rev Biochem* 54: 729–764.

Niederwieser A, Staudenmann W, Wetzel E (1982a) Automated HPLC of pterins with or without column switching. In Wachter H, Curtius HCh, Pfleiderer W, eds. *Biochemical and Clinical Aspects of Pteridines,* Vol. 1 Berlin: de Gruyter, 81–102.

Niederwieser A, Curtius H-Ch, Wang M, Leupold D (1982b) Atypical phenylketonuria with defective biopterin metabolism. Monotherapy with tetrahydrobiopterin or sepiapterin, screening and study of biosynthesis in man. *Eur J Pediatr* 138: 110–112.

O'Kusky JR, Boyes BE, Walker DG, McGeer EG (1991) Cytomegalovirus infection of the developing brain alters catecholamine and indoleamine metabolism. *Brain Res* 559: 322–330.

Ponzone A, Blau N, Guardamagna O, Ferrero GB, Dianzani I, Endres W (1990) Progression of 6-pyruvoyltetrahydropterin synthase deficiency from a peripheral into a central phenotype. *J Inher Metab Dis* 13: 298–290.

Robertson D, Goldberg MR, Onrot J et al (1986) Isolated failure of autonomic noradrenergic neurotransmission. *N Engl J Med* 314: 1494–1497.

Scriver CR, Kaufman S, Woo SLC (1990) The Hyperphenylalaninemias. In Scriver CR, Beaudet AL, Sly WS, Valle D, eds. *The Metabolic Basis of Inherited Disease,* 6th edn. New York: McGraw-Hill, 495–546.

Sharpless NS, McCann DS (1971) Dopa and 3-O-methyldopa in cerebrospinal fluid of Parkinsonian patients during treatment with oral L-dopa. *Clin Chim Acta* 31: 155–169.

Shaywitz BA, Cohen DJ, Bowers MB (1975) Reduced cerebrospinal fluid 5-hydroxyindoleacetic acid and homovanillic acid in children with epilepsy. *Neurology* 25: 72–76.

Silverstein F, Johnston MV (1984) Cerebrospinal fluid monoamine metabolites in infant spasms. *Neurology* 34: 102–105.

Smith I (1990) Disorders of tetrahydrobiopterin metabolism. In Fernandes J, Saudubray JM, Tada K, eds. *Inborn Metabolic Disease.* Heidelberg: Springer Verlag, 183–197.

Smith I, Dhondt JL (1985) Birthweight in patients with defective biopterin synthesis. *Lancet* 1: 818.

Smith I, Clayton BE, Wolff OH (1975) New variant of phenylketonuria with progressive neurological illness unresponsive to phenylalanine restriction. *Lancet* 1: 1108–1111.

Smith I, Hyland K, Kendall B, Leeming R (1985) Clinical role of pteridine therapy in tetrahydrobiopterin deficiency. *J Inher Metab Dis* 8 (Suppl 1): 39–45.

Sugie H, Sugi Y, Kato N, Fukuyama Y (1989) A patient with infantile spasms and low homovanillic acid levels in cerebrospinal fluid: L-dopa dependent seizures? *Eur J Pediatr* 148: 667–668.

Surtees R, Hyland K (1989) A method for the measurement of *S*-adenosylmethionine in small volume samples of cerebrospinal fluid or brain using high performance liquid chromatography. *Anal Biochem* **181**: 331–335.

Wenk GL, Naidu S, Casanova MF, Kitt CA, Moser H (1991) Altered neurochemical markers in Rett's syndrome. *Neurology* **41**: 1753–1756.

Wester P, Bergstromm U, Eriksson A, Gezelius C, Hardy J, Winblad G (1990) Ventricular cerebrospinal fluid monoamine transmitter and metabolite concentrations reflect human brain neurochemistry in autopsy cases. *J Neurochem* **54**: 1148–1156.

J. Inher. Metab. Dis. 16 (1993) 691–703
© SSIEM and Kluwer Academic Publishers. Printed in the Netherlands

Non-ketotic Hyperglycinaemia: Molecular Lesion, Diagnosis and Pathophysiology

K. TADA and S. KURE
Department of Pediatrics and Biochemical Genetics, Tohoku University School of Medicine, Seiryo-machi 1-1, Aoba-ku, Sendai 980, Sendai, Japan

Summary: Non-ketotic hyperglycinaemia (NKH) is a well-recognized metabolic cause of life-threatening illness in the neonate. The fundamental defect is in the glycine cleavage system, which consists of four protein components. Our study revealed that the majority of NKH patients had a specific defect in P-protein (glycine decarboxylase). The primary lesion of NKH at gene level was investigated, using cDNA encoding human glycine decarboxylase. A three-base deletion resulting in deletion of Phe^{756} was found in a Japanese patient with NKH. The majority of NKH patients in Finland, where there is a high incidence of NKH, were found to be due to a common mutation, a point mutation resulting in the amino acid substitution of Ile^{564} for Ser^{564}. Prenatal diagnosis is feasible by determining the activity of the glycine cleavage system and is also possible by DNA analysis. Recent findings suggest that a high concentration of glycine in the brain may contribute to the pathophysiology of NKH by overactivating N-methyl-D-aspartate receptors allosterically, which may result in intracellular calcium accumulation, DNA fragmentation and neuronal death. These provide the possibility that early treatment with N-methyl-D-aspartate receptor antagonist may prevent brain damage in NKH.

Non-ketotic hyperglycinaemia (NKH, McKusick 238300) is an autosomal recessive disorder characterized by abnormally high concentrations of glycine in plasma and cerebrospinal fluid. It is a well-recognized metabolic cause of life-threatening illness in the neonate. NKH is classified in two types from clinical aspects: neonatal type and late-onset type (Nyhan 1989; Tada 1987, 1990; Tada and Hayasaka 1987). The neonatal type, which is more severe and common, is characterized by rapid development of neurological symptoms such as lethargy, muscular hypotonia, apnoea and seizures in the newborn period. Most patients die within a few weeks of life, whereas the survivors show severe psychomotor retardation. In the late-onset type the patients develop neurological symptoms in a variety of degrees after the neonatal period. According to our experience of 30 cases of NKH, 26 cases (87%) were of the neonatal type. Among them, 22 cases died between 6 days and 5 years of life. The remaining 4 cases survived but were severely retarded (Tada 1987, 1991).

The prevalence of NKH is unknown because many patients must die in early infancy undiagnosed. In northern Finland the prevalence is estimated to be 1 : 12 000 (Von Wendt et al 1979).

PRIMARY LESION AT THE PROTEIN LEVEL

The fundamental defect of NKH is in the glycine cleavage system (EC 2.1.2.10), which was demonstrated by Tada et al in 1969. The glycine cleavage system is composed of four protein components: P-protein (a pyridoxal phosphate-dependent glycine decarboxylase), H-protein (a lipoic acid-containing protein), T-protein (a tetrahydro-folate-requiring enzyme) and L-protein (lipoamide dehydrogenase) (Kikuchi 1973).

We have so far analysed glycine cleavage activity and its component proteins in the livers from 30 patients with NKH (Tada 1987, 1991; Hayasaka et al 1983). The glycine cleavage activity was undetectable or extremely low in the neonatal type, whereas the late-onset type showed some residual activity. Thus the clinical phenotypes do seem to relate to the degree of the defect in glycine cleavage activity. Analysis of component proteins of the glycine cleavage system showed that the majority of NKH patients (26 out of 30, 87%) had a specific defect in P-protein and the remaining a specific defect in T-protein. The component analysis was made in the brain from seven autopsied cases. The sites of defect in these cases are identical in both the brain and liver.

A NOVEL METHOD OF ENZYMATIC DIAGNOSIS OF NKH

The glycine cleavage system is specifically expressed in liver, kidney and brain, but not in easily-obtainable tissues such as fibroblasts or leukoytes. Liver biopsy is, therefore, necessary for the enzymatic diagnosis of NKH. Because of the difficulty of obtaining liver biopsy from healthy subjects, there has been no information about glycine cleavage system activities of healthy carriers for NKH.

Recently Kure and colleagues (1992a) developed a new assay method for the glycine cleavage system using blood samples instead of biopsied liver samples. The new method is based on their observation that the glycine cleavage system is induced in B-lymphocytes by infection and transformation using Epstein–Barr virus (EBV). It was found that the overall glycine cleavage system and P-protein assays in transformed lymphoblasts are reliable for evaluating the glycine cleavage system activity of patients. The reliability of the lymphoblast assay is supported by three lines of evidence. First, the results of overall glycine cleavage system assay and P-protein assay of lymphoblasts from six patients with NKH agreed well with those of their liver biopsy samples. Second, the mean overall glycine cleavage system and P-protein activities in parents were 38% and 45%, respectively, of those in control subjects, which was consistent with the expected result for heterozygotic carriers of NKH, an autosomal recessive disorder. Third, no lymphoblasts from normal subjects had enzyme activity lower than those from obligate carriers or patients with NKH. This method provides the following advantages:

(1) Enzymatic diagnosis of NKH is feasible using the peripheral blood, instead of biopsied liver tissue.
(2) Differential diagnosis between NKH and ketotic hyperglycinaemia is easily made.

(3) Carrier detection is possible using the peripheral blood.
(4) Structural analysis of mRNA encoding the glycine cleavage system, especially P-protein, is easily applicable using lymphoblasts from NKH patients.

PRIMARY LESION AT THE GENE LEVEL

Since the majority of NKH is caused by a specific defect in P-protein (glycine decarboxylase), we cloned cDNA encoding human glycine decarboxylase from human placenta gt11 expression library, using the specific antibody against rat P-protein (Kure et al 1991a). This clone was 3705 bp in length and encoded 1020 amino acids. The authenticity of this cDNA was verified by concordance with amino acid sequence of P-protein that was partially known and also by expression study in Cos7 cells. Using this cDNA as a hybridization probe, the primary lesion of NKH at gene level was investigated.

In a Japanese patient with neonatal type NKH, who had been proved to have a specific defect in P-protein, mRNA was isolated from the liver obtained at autopsy at 4 years of age. Then the sequence of P-protein cDNA synthesized from the patient's mRNA was determined (Kure et al 1991a). A three-base deletion that resulted in deletion of Phe[756] was found (Figure 1). In order to confirm the pathogenicity of this mutation, we expressed the normal and mutant P-protein cDNA in Cos7 cells and determined the P-protein activity in those cells. As shown in Figure 2, Cos7 cells in which normal P-protein cDNA was expressed showed an activity of 6.9 nmol/mg protein per hour, which was almost equivalent to that of human liver. In contrast, Cos7 cells in which the mutant cDNA was expressed showed no activity, indicating that the three-base deletion could cause NKH.

It is known that the incidence of NKH is high in Finland. Therefore, we investigated the mutation site in Finnish patients, using lymphoblast cells established by EB virus

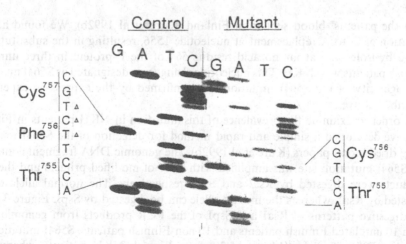

Figure 1 Sequence analysis of the three-base deletion in the P-protein cDNA from an NKH patient. The three-base deletion (TCT) resulted in the deletion of phenylalanine at 756. (Kure et al 1991a)

Expression of normal and mutant P-protein cDNA in Cos7 cell

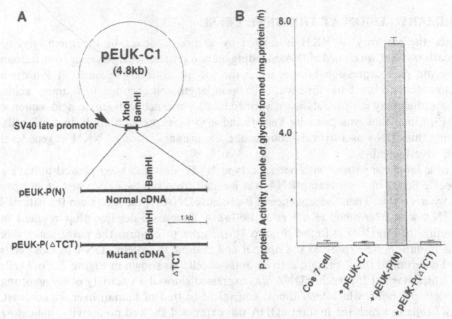

Figure 2 (A) Construction of the expression vectors containing normal and mutant P-protein cDNA. (B) P-protein activity of Cos7 cells expressing normal and mutant P-protein cDNA. Recombinant plasmid (20 μg) was transfected into Cos7 cells by using lipofection reagent. After 48 h incubation, the P-protein activity in the cells was assayed. (Kure et al 1991a)

from the patients' blood sent from Finland (Kure et al 1992b). We found a point mutation of G to A replacement at nucleotide 1556, resulting in the substitution of serine by isoleucine at amino acid residue 564 of the P-protein in three unrelated Finnish patients with NKH. This nucleotide change was designated as S564I mutation. Pathogenicity of the S564I mutation was confirmed by the expression experiment described above.

In order to examine the prevalence of this mutation in NKH patients in Finland, we have developed a simple and rapid method for detection of the S564I mutation using dried blood papers (Kure et al 1992b). The genomic DNA fragment containing the S564I mutation site was amplified with a set of modified primers and the PCR products were digested by RsaI and SspI, respectively. The normal allele can be digested by RsaI, whereas the mutant allele can be digested by SspI. Figure 3 shows the digestive patterns of RsaI and SspI of the PCR products from genomic DNA from 10 unrelated Finnish patients and 10 non-Finnish patients. S564I mutation was found in 70% of alleles (14 out of 20) from Finnish NKH patients, whereas this mutation was not detected in 10 non-Finnish patients with NKH. These observations suggest that the S564I mutation had been distributed by a founder effect, not because

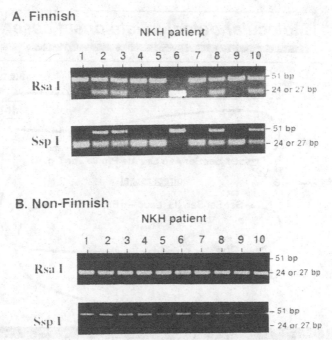

Figure 3 Detection of the point mutation in DNA from Finnish and non-Finnish patients with NKH. The photomicrographs show the digestion patterns of RsaI and SspI of the PCR-amplified product from genomic DNA of (A) 10 Finnish patients with NKH and (B) 10 non-Finnish patients with NKH. (Kure et al 1992b)

this is a hot spot of mutation.

The amino acid (Ser[564]) altered by the S564I mutation is present in a particular domain of the P-protein. Ser[564] is located at the NH_2-terminus in a decapeptide (Ser[564] to Trp[573]) that is directly repeated in a nonapeptide from Ser[807] to Trp[815] (Figure 4). The amino acid residue that binds to pyridoxal phosphate, a cofactor of P-protein, was determined in chicken P-protein (Fujiwara et al 1987) and corresponds to Lys[754] of the human P-protein based on the structural similarity between the chicken and human enzymes. This critical residue resides between the two repetitive structures in P-protein polypeptide, suggesting that a region from Ser[564] to Trp[815] is important in P-protein function. In fact, the mutant P-protein polypeptide identified in the Japanese patient described above also had a defect in the region from Ser[564] to Trp[815]: Phe[756] was deleted by three-base deletion in the P-protein gene. Because Phe[756] is located closely to Lys[754], the binding site of a pyridoxal phosphate, the deletion of Phe[756] seems likely to interfere with its binding or its function (Figure 4). Site-directed mutagenesis of P-protein might elucidate the structure–function relationship of this protein.

The neonatal type of NKH is an overwhelming illness with undetectable activity of the glycine cleavage system, whereas the late-onset type is milder in clinical course with some residual activity of the glycine cleavage system. Recently we found a

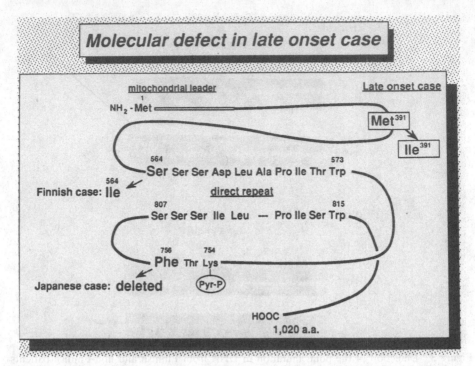

Figure 4 Molecular defect in NKH. The decapeptide between Ser[564] and Try[573] is almost directly repeated in the nonapeptide between Ser[807] and Try[815]. A common mutation, S564I, in Finnish NKH patients with neonatal type and a mutation in a Japanese NKH patient of neonatal type were found in this region. The mutation-site in the NKH patient with late-onset type was found far from the region

missense mutation in a patient with late-onset type of NKH who showed some residual activity of P-protein. The mutation causes amino acid replacement of methionine by isoleucine at amino acid 391, which is far from the mutation-sites found in the neonatal type (Figure 4).

PRENATAL DIAGNOSIS

There is a special demand for prenatal diagnosis, since no effective treatment is available for NKH. Prenatal diagnosis using cultured amniotic cells is not possible, because the glycine cleavage system is not manifested in cultured amniotic fluid cells. In 1987, we found the existence of glycine cleavage system in chorionic villi of placenta (based on a retrospective study) and suggested the feasibility of prenatal diagnosis of NKH by chorionic villus sampling (Hayasaka et al 1987). Subsequently we have had opportunities to attempt prenatal diagnosis of NKH using chorionic villi (Hayasaka et al 1990). Thirty-one pregnancies of women who had previous children affected with NKH of neonatal type were monitored. Chorionic villus sampling was performed between the 8th and 16th weeks of gestation. The biopsied chorionic villi were immediately frozen and stored at −80°C until the analyses. In 23 out of 31 cases, glycine cleavage system activity was found to be normal. These pregnancies continued and healthy babies were born after full-term pregnancies. The plasma glycine levels of the babies were within normal limits. In the remaining 8 cases, glycine cleavage system activity was undetectable, suggesting that the fetus was affected with NKH. The pregnancies were terminated at the parents request. The glycine cleavage system activity in the liver and brain from aborted fetuses was nearly undetectable. Thus the fetuses were confirmed to be affected with NKH.

These findings clearly indicate that prenatal diagnosis of NKH is possible by determining the glycine cleavage system activity in chorionic villi. DNA diagnosis also is possible when the mutant site of the family is known. We have made a prenatal diagnosis in a Finnish family by the method of detecting the S564I mutation described previously (Figure 5). The first child of this family was affected with NKH and proved to be homozygous for S564I mutation. The parents were heterozygous for the mutation. The chorionic villi at the 12th week of gestation of the second pregnancy of the mother were examined and found to be homozygous for the mutation. The glycine cleavage activity was undetectable in addition. Therefore, the fetus was diagnosed as 'affected'. This was confirmed on DNA examination of the tissue of the aborted fetus.

PATHOPHYSIOLOGY

A marked elevation of glycine concentration in CSF is characteristic of NKH and indicates an accumulation of glycine in the brain due to a defect of the glycine cleavage system. Actually a high concentration of glycine was reported in the brain of patients with NKH (Perry et al 1975; Tada 1988). Although a high concentration of glycine in the brain is presumed to be responsible for neurological impairment in NKH, its detailed pathophysiology has been unclear. However, several recent

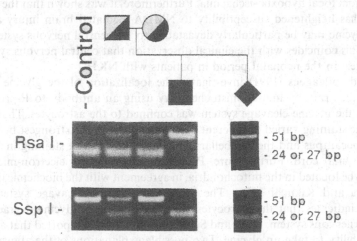

Figure 5 Prenatal diagnosis of non-ketotic hyperglycinaemia by DNA analysis. The modified PCR method was followed by RsaI and SspI digestions. The digested DNA fragments were size-separated by 5.5% agarose gel electrophoresis and visualized by ethidium bromide staining (Kure et al 1992b)

J. Inher. Metab. Dis. 16 (1993)

studies have provided a greater understanding of the relationship of glycine and neuroexcitotoxicity. Glycine has so far been believed to act as an inhibitory neurotransmitter at a strychinine-sensitive receptor. Recently it was discovered by Johnson and Ascher (1987) that glycine has an excitatory property that potentiates one of the glutaminergic receptors, called an N-methyl-D-aspartate (NMDA) receptor. They suggested that glycine enhances NMDA-mediated responses of a site closely associated with the NMDA receptor, based on experimental results using cultured mouse brain neurons. Larson and Beitz (1988) demonstrated that glycine administration enhanced NMDA-induced seizures in mice whose classic glycine receptor had been blocked with strychinine. This finding indicated that glycine is a harmful enhancer of NMDA-response *in vivo*. It is known that classic glycine receptors exist mainly in the brain stem and spinal cord (Young and Snyder 1973; Zarbin et al 1981), whereas NMDA receptors are located in the whole brain (Monaghan and Cotman 1985). According to Daly and colleagues' study using rats (1976), the glycine concentration of the brain is high in the region where glycine cleavage system activity is low (brain stem and spinal cord), and low in the region where glycine cleavage system activity is high (telencephalone and cerebellum). These findings suggest that a high concentration of glycine may be more harmful to the region where NMDA receptors are distributed than to the region where strychinine-sensitive receptors are located.

Neurophysiological evidence has accumulated indicating that excessive activation of excitatory amino acid receptors, particularly NMDA receptor, has been implicated in the pathogenesis of neuronal injury in a number of neurological disorders including hypoxia–ischaemia, hypoglycaemia or physical brain trauma (McDonald and Johnston 1990). Several groups have demonstrated that the administration of selective antagonists of the NMDA receptor channel reduces the severity of brain injury resulting from focal hypoxia–ischaemia. Furthermore, it was shown that the developing brain has heightened susceptibility to NMDA-mediated brain injury and high levels of glycine may be particularly devastating to the central nervous system of the neonate. This coincides with the clinical observation that central nervous syndromes occur acutely in the neonatal period in patients with NKH.

Sato and colleagues (1991) investigated the localization of the glycine cleavage system in rat brain by immunohistochemistry using an antibody to P-protein and found that the glycine cleavage system was confined to the astrocytes. The intensity of astrocyte staining varied in different brain regions, with the strongest being noted in the hippocampus and the cerebellar cortex and the weakest staining in the brain stem and spinal cord. Furthermore, P-protein was found by electron-microscopic analysis to be located in the mitochondria, in agreement with the biochemical findings (Motokawa and Kikuchi 1971). The presence of glycine cleavage system in the astrocytes indicates that the astrocytes are the main site of glycine degradation in the central nervous system. Hosli and Schousboe (1986) have reported that astrocytes have the ability to take up glycine. Two possible explanations of the above findings are that the overflow of glycine released into the synaptic cleft from presynaptic sites is taken up by the astrocytes or that excess glycine in the neurons is transported to the astrocytes and degraded. Glycine cleavage system in astrocytes may be closely related to NMDA receptors, because there is a correlation between the sites of the glycine cleavage system and those of NMDA receptors. In this respect, it is worthy of note that the regional distribution of glutamine synthase (Norenberg 1979), which is located in astrocytes and plays an important role in inactivation of the excitatory property of glutamic acid, coincides well with that of P-protein.

Using the above findings, the following hypothesis can be made to explain the pathophysiology of NKH (Figure 6). A defect of glycine cleavage system (GCS) in the brain (astrocytes) brings about a rise of glycine concentration in synaptic cleft. An excess of glycine allosterically overactivates NMDA receptors, which may result in intracellular calcium accumulation, seizures and neuronal injury, as was suggested from the neurotoxicity of excitatory amino acids in the pathophysiology of developmental neurological disorders (McDonald and Johnston 1990). However, the subcellular mechanism of neuronal cell death by excitotoxicity still remains to be resolved.

Recently we found an interesting phenomenon in glutamate neurotoxicity (Kure et al 1991b). The cerebral neocortices were removed from Wistar rats (15–16 day gestation) and subjected to dissociated neuron culture. It was found that chromosomal DNA of the cultured neurons was degraded into nucleosomal-size DNA fragments by the addition of glutamate, prior to the glutamate-induced neuronal death. Both the neuronal death and DNA fragmentation were prevented by inhibitors of endonuclease; implying that the DNA fragmentation is caused by activated endonuclease. The activation mechanism of endonuclease is unknown. However, increased

A Defect of GCS in Brain

⇩

Elevation of Glycine in Brain

⇩

Overstimulation of NMDA Receptors

(at a glycine modulatory site)

⇩

Cation Channel Activation

⇩

Intracellular Ca Accumulation

⇩

Activation of Endonuclease

⇩

DNA Fragmentation

⇩

Neuronal Death

Figure 6   A hypothesis for the pathophysiology of NKH

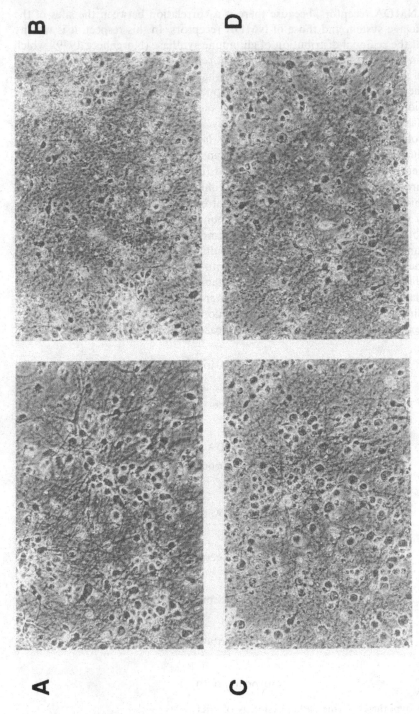

Figure 7 The neurotoxicity of NMDA and its enhancement by glycine. The dissociated neuron cultures were established from the cerebral neocortices of fetal Wistar rats, and maintained in Dulbecco's modified Eagle medium supplemented with 10% fetal bovine serum at 37°C for 10 days. The phase-constant photomicrographs were taken after treatments for 24 h with 500 μmol/L NMDA (B), 50 μmol/L NMDA (C), and 50 μmol/L NMDA plus 300 μmol/L glycine (D). (A) shows a control with no treatment

intracellular Ca^{2+} concentration seems to be important, as evidenced by the fact that the Ca^{2+} ionophore A23187, instead of glutamate, could induce DNA fragmentation (Kure et al 1991b). Moreover, an injection of glutamate into the rat hippocampi resulted in DNA fragmentation with a similar time course to that observed in neuronal death *in vitro*.

Based on the above findings, we consider that DNA fragmentation by glutamate may be the subcellular mechanism of neuroexcitotoxicity. Furthermore, severe central nervous symptoms and the grave prognosis seen in NKH are presumed to be due to a similar mechanism through the overactivation of NMDA receptors by glycine accumulated in the brain. This view is supported by our experimental findings in the cultured neuron as shown in Figure 7. The addition of 300 μmol/L glycine, comparable to the level in the brain of NKH patients, to the cultured neurons caused neuronal death in the presence of 50 μmol/L NMDA, whereas neurons remained intact in the presence of 50 μmol/L NMDA alone (Figure 7). Recently, tentative treatment using NMDA antagonists has been tried in NKH. Ohya and colleagues (1991) reported that the administration of ketamine, a NMDA receptor antagonist, to a patient with NKH at 7 months of age brought a partial improvement of neurological symptoms and EEG findings.

The Baltimore group (Homosb et al 1991) made a preliminary report of combined therapy with benzoate and dextromethorphan, an NMDA antagonist, in an infant with NKH and speculated that dextromethorphan may have improved his outcome. We also made a trial of oral administration of dextromethorphan in a patient with NKH at the age of 3 months. Surprisingly, his intractable seizures almost disappeared, with a remarkable improvement of EEG findings.

Clinical observations indicate that brain damage in NKH occurs in the early stages of life. Therefore, treatment during the early stage of life is thought to be essential to prevent irreversible brain damage. There is a possibility that early treatment with NMDA receptor antagonist may prevent brain damage in NKH, although further experience is needed.

ACKNOWLEDGEMENT

This work was supported by Grant-in-Aid for Scientific Research from the Ministry of Education, Science and Culture, and by Grants from the Ministry of Health and Public Welfare, Japan.

REFERENCES

Daly EC, Nadi NS, Aprison MH (1976) Regional distribution and properties of the glycine cleavage system within the central nervous system of the rat. *J Neurochem* **26**: 179–185.

Fujiwara K, Okamura-Ikeda K, Motokawa Y (1987) Amino acid sequence of the phosphopyridoxyl peptide from P-protein of the chicken liver glycine cleavage system. *Biochem Biophys Res Commun* **149**: 621–627.

Hayasaka K, Tada K, Kikuchi G, Winter S, Nyhan WL (1983) Nonketotic hyperglycinemia: two patients with primary defects of P-protein and T-protein, respectively, in the glycine cleavage system. *Pediatr Res* **17**: 926–970.

Hayasaka K, Tada K, Fueki N et al (1987) Feasibility of prenatal diagnosis of nonketotic hyperglycinemia: existence of the glycine cleavage system in placenta. *J Pediatr* **110**: 124–126.

Hayasaka K, Tada K, Fueki N, Aikawa J (1990) Prenatal diagnosis of nonketotic hyperglycinemia: Enzymatic analysis of the glycine cleavage system in chorionic villi. *J Pediatr* **116**: 444–445.

Homosb A, Johnston MV, McDonald JW, Francamano C, Niedermeyer E, Valle D (1991) One-year experience with combination benzoate and excitatory amino acid antagonist therapy for nonketotic hyperglycinemia (abstract of 20th Annual Meeting of Child Neurology Society). *Ann Neurol* **30**: 469.

Hosli EHL, Schousboe A (1986) Amino acid uptake. In Sergy F, Antonia V, eds. *Astrocytes: Biochemistry, Physiology, and Pharmacology of Astrocytes*, Vol. 2. Orlando: Academic Press, 133–153.

Johnson JW, Ascher P (1987) Glycine potentiates the NMDA response in cultured mouse brain neurons. *Nature (London)* **325**: 529–531.

Kikuchi G (1973) The glycine cleavage system: Composition, reaction mechanism, and physiological significance. *Mol Cell Biochem* **1**: 169–187.

Kure S, Narisawa K, Tada K (1991a) Structural and expression analyses of normal and mutant mRNA encoding glycine decarboxylase: Three base deletion in mRNA causes nonketotic hyperglycinemia. *Biochem Biophys Res Commun* **174**: 1176–1182.

Kure S, Tominaga T, Yoshimoto T, Tada K, Narisawa K (1991b) Glutamate triggers internucleosomal DNA cleavage in neuronal cells. *Biochem Biophys Res Commun* **179**: 39–45.

Kure S, Narisawa K, Tada K (1992a) Enzymatic diagnosis of nonketotic hyperglycinemia with lymphoblasts. *J Pediatr* **120**: 95–98.

Kure S, Takayanagi M, Narisawa K, Tada K, Leisti J (1992b) Identification of a common mutation in Finnish patients with nonketotic hyperglycinemia. *J Clin Invest* **90**: 160–164.

Larson AA, Beitz AJ (1988) Glycine potentiates strychnine induced convulsions: role of NMDA receptors. *J Neurosci* **8**: 3822–3826.

McDonald JW, Johnston MV (1990) Physiological and pathophysiological roles of excitatory amino acids during central nervous system development. *Brain Res Rev* **15**: 41–70.

Monaghan DT, Cotman CW (1985) Distribution of N-methyl-D-aspartate-sensitive L-[^{3}H]glutamate binding sites in rat brain. *J Neurosci* **5**: 2909–2919.

Motokawa Y, Kikuchi G (1971) Glycine metabolism in rat liver mitochondria. Intramitochondrial localization of the reversible glycine cleavage system and serine hydroxymethyltransferase. *Arch Biochem Biophys* **146**: 461–466.

Norenberg MD (1979) The distribution of glutamine synthase in the rat central nervous system. *J Histochem Cytochem* **27**: 756–762.

Nyhan WL (1989) Nonketotic hyperglycinemia. In Scriver CR, Beaudet AL, Sly WS, Valle D, eds. *The Metabolic Basis of Inherited Disease*, 6th edn. New York: McGraw-Hill, 743–753.

Ohya Y, Ochi N, Mizutani N, Hayakawa C, Watanabe K (1991) Nonketotic hyperglycinemia: Treatment with NMDA antagonist and consideration of neuropathogenesis. *Pediatr Neurol* **7**: 65–68.

Perry TL, Urquhart N, McLean J et al (1975) Nonketotic hyperglycinemia. *N Engl J Med* **292**: 1269–1273.

Sato K, Yoshida S, Fujiwara K, Tada K, Tohyama M (1991) Glycine cleavage system in astrocytes. *Brain Res* **567**: 64–70.

Tada K (1987) Nonketotic hyperglycinemia: Clinical and metabolic aspects. *Enzyme* **38**: 27–35.

Tada K (1988) Pathogenesis and prenatal diagnosis of nonketotic hyperglycinemia. *Nipponrinsho* **46**: 1865–1877.

Tada K (1990) Nonketotic hyperglycinemia. In Fernandes J, Saudubray JM, Tada K, eds. *Inborn Metabolic Diseases*. Heidelberg: Springer-Verlag, 323–329.

Tada K (1991) Nonketotic hyperglycinemia: Pathogenesis and diagnosis. *Nihon Sententaishaijo Gakkai Zasshi* 7: 1–13.

Tada K (1992) Nonketotic hyperglycinemia: A life-threatening disorder in the neonate. *Early Human Dev* 29 75–81.

Tada K, Hayasaka (1987) Nonketotic hyperglycinemia: Clinical and biochemical aspects. *Eur J Pediatr* 146: 221–227.

Tada K, Narisawa K, Yoshida T et al (1969) Hyperglycinemia: a defect in glycine cleavage reaction. *Tohoku J Exp Med* 98: 289–296.

Von Wendt L, Hirvasniemi A, Simila S (1979) Nonketotic hyperglycemia: A genetic study of 13 Finnish families. *Clin Genet* 15: 411–417.

Young AB, Snyder SH (1973) Strychinine binding associated with glycine receptors of the central nervous system. *Proc Natl Acad Sci USA* 70: 2832–2836.

Zarbin MA, Wamsley JK, Kuhar MJ (1981) Glycine receptor: Light microscopic autoradiographic localization with [^{3}H]strychinine. *J Neurosci* 1: 532–547.

J. Inher. Metab. Dis. 16 (1993) 704–715
© SSIEM and Kluwer Academic Publishers. Printed in the Netherlands

Inherited Disorders of GABA Metabolism

C. Jakobs[1], J. Jaeken[2] and K. M. Gibson[3]

[1]*Department of Pediatrics, Free University Hospital, de Boelelaan 1117, 1081 HV Amsterdam, The Netherlands;* [2]*Department of Pediatrics, University Hospital Gasthuisberg, 49 Herestraat, B-3000 Leuven, Belgium;* [3]*Metabolic Disease Center, Baylor Research Institute and Baylor University Medical Center, 3500 Gaston Avenue, Dallas, Texas, USA*

Summary: Gamma-aminobutyric acid (GABA), a major inhibitory neurotransmitter in the mammalian central nervous system, is produced from glutamic acid in a reaction catalysed by glutamic acid decarboxylase. The sequential actions of GABA-transaminase (converting GABA to succinic semialdehyde) and succinic semialdehyde dehydrogenase (oxidizing succinic semialdehyde to succinic acid) allow oxidative metabolism of GABA through the tricarboxylic acid cycle. The inherited disorders of GABA metabolism include: (1) pyridoxine-dependent seizures (?glutamic acid decarboxylase deficiency) (> 50 patients); (2) GABA-transaminase deficiency (2 patients/1 family); (3) succinic semialdehyde dehydrogenase deficiency (32 patients/21 families); and (4) homocarnosinosis associated with serum carnosinase deficiency (3 patients/1 family). Homocarnosine is a brain-specific dipeptide of GABA and L-histidine. Of these four defects, definitive enzymatic diagnoses have been made only for GABA-transaminase and succinic semialdehyde dehydrogenase deficiencies. The presumptive mode of inheritance for all disorders is autosomal recessive, and all are associated with central nervous system dysfunction. Only succinic semialdehyde dehydrogenase deficiency manifests organic aciduria, which may account for the higher number of patients identified with this disorder; identification of additional patients with some of the other disorders will require increased request for analysis of cerebrospinal fluid metabolites by paediatricians and neurometabolic specialists.

The inherited disorders of GABA metabolism (Figure 1) have been reviewed by several investigators (Gibson et al 1986; Scriver and Perry 1989; Jaeken et al 1990; Jaeken 1990). Subsequently, the only additional patients identified have been those with pyridoxine-dependent seizures and succinic semialdehyde dehydrogenase deficiency. The reason for this may lie in the nature of the disorders. Pyridoxine-dependent seizures are diagnosed by response to pyridoxine therapy. GABA-transaminase deficiency and homocarnosinosis are diseases in which diagnostic metabolites accumulate predominantly in cerebrospinal fluid (CSF), and to a smaller extent in plasma. Succinic semialdehyde dehydrogenase deficiency, however, presents with 4-hydroxybutyric aciduria due to conversion of accumulated succinic semi-

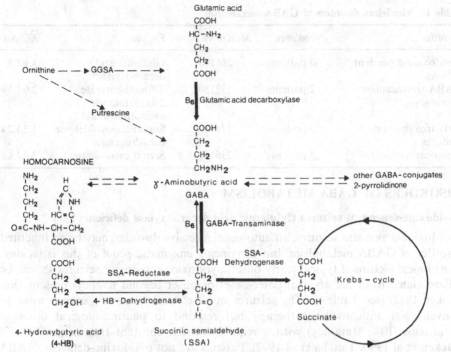

Figure 1 Metabolic interrelations of gamma-aminobutyric acid (GABA) metabolism. GGSA = gamma-glutamate semialdehyde dehydrogenase. Heavy arrows indicate major direction of metabolism

aldehyde (Figure 1) to 4-hydroxybutyric acid, and the organic aciduria may be detected by routine organic acid analysis.

Although its primary metabolic role is that of neurotransmitter in the central nervous system, GABA is by no means localized only in brain. GABA is found in other non-neural tissues, including pancreas and kidney (Erdo 1985). GABA is produced from putrescine as well as glutamic acid, although the latter is the major source of GABA. Since GABA is converted to succinic acid, it is capable of supporting oxidative metabolism in neural and non-neural tissues. The role of GABA as a neurotransmitter as well as its role in mammalian non-neural tissue have been reviewed in detail (Iversen 1982; Johnston and Singer 1982; Erdo 1985).

This review describes inherited disorders involving the metabolic pathway of GABA, and the metabolism of the GABA-containing dipeptide homocarnosine (see Table 1). The disorders discussed will include pyridoxine-dependent seizures (?GAD deficiency; McKusick 266100; EC 4.1.1.15); GABA-transaminase deficiency (McKusick 137150; EC 2.6.1.19); succinic semialdehyde dehydrogenase deficiency (McKusick 271980; EC 1.2.1.24), and homocarnosinosis associated with serum carnosinase deficiency (McKusick 236130; EC 3.4.13.3). Clinical, metabolic, enzymatic and pertinent molecular aspects of these four disorders are presented.

J. Inher. Metab. Dis. 16 (1993)

Table 1   Mendelian disorders of GABA metabolism

Disorder	Prevalence	McKusick No.	Enzyme	EC No.
Pyridoxine-dependent seizures	> 50 patients	266100	Glutamic acid decarboxylase?	4.1.1.5
GABA-transaminase deficiency	2 patients	137150	4-Aminobutyrate 2-ketoglutarate aminotransferase	2.6.1.19
4-Hydroxybutyric aciduria	32 patients	271980	Succinic semialdehyde dehydrogenase	1.2.1.24
Homocarnosinosis	3 patients	236130	Serum carnosinase	3.4.13.3

DISORDERS OF GABA METABOLISM

Pyridoxine-dependent seizures (?glutamic acid decarboxylase deficiency)

Pyridoxine-dependent seizures, an autosomal recessive disorder, may be an inherited disorder of GABA metabolism, but convincing enzymatic proof of this is lacking. The clinical picture of typical (early-onset) pyridoxine-responsive seizures should be differentiated from the atypical (later-onset or onset beyond infancy) presentation (Coker 1992) (see Table 2). The seizures in all clinical forms are unresponsive to conventional anticonvulsant therapy but respond to pharmacological doses of pyridoxine (10–100 mg/day) with a requirement for constant treatment (Table 2) (Jaeken et al 1990; Tanaka et al 1992). Patients are not pyridoxine-deficient. GABA concentration in brain has been measured (*post mortem*) in one patient and CSF GABA in another; values were low in both. No data are available on CSF homocarnosine concentrations. In one patient GABA concentrations in CSF increased and seizures resolved following institution of pyridoxine therapy. Glutamic acid decarboxylase (GAD), the enzyme responsible for GABA production, requires pyridoxal-5'-phosphate as cofactor for the decarboxylation reaction (Figure 1). It has been widely assumed that patients with pyridoxine-responsive seizures have defective GAD activity due to abnormal cofactor binding.

Yoshida and coworkers (1971) presented the only evidence that pyridoxine-dependent seizures are associated with GAD deficiency. These investigators documented an abnormality of cofactor (pyridoxal-5'-phosphate) binding to apo-GAD in biopsied renal cortex from a patient with mental retardation and pyridoxine-dependent seizures. Using uniformly ^{14}C-radiolabelled glutamate, no evolution of [14]CO_2 was detected without added pyridoxal-5'-phosphate in the *in vitro* assay, but [14]CO_2 evolution in the patient's kidney extract was stimulated above the control range when cofactor was added. These results are of interest, but may be inconclusive. Recent work has demonstrated different molecular forms of GAD (molecular weights 65 and 67 kDa, respectively) (Bu et al 1992), and both GAD forms may be found in brain and pancreas (Baekkeskov et al 1990). Abnormalities of the 65 kDa form of GAD may be associated with insulin-dependent diabetes mellitus (IDDM) and stiff-man syndrome, a rare neurological disease associated with IDDM (Baekkeskov et al 1990). With respect to the findings of Yoshida and colleagues using kidney as the source of GAD for enzyme study, we do not know which molecular weight species

Table 2 Pyridoxine-dependent seizures – clinical and biochemical findings and treatment

Clinical presentation

Typical
    Onset of convulsions before or shortly after birth
    Rapid response to pyridoxine
    Refractoriness to other anticonvulsants
    Dependence on continued therapy
    Absence of pyridoxine deficiency

Atypical
    Later onset of seizures
    Prolonged seizure-free intervals without pyridoxine (as long as 5 months)
    The need for larger pyridoxine doses in some patients
    Higher incidence

Biochemical findings

Brain GABA (1 patient *post mortem*) and CSF GABA (1 patient) were low
No data available on CSF homocarnosine concentrations
Hypothesis: genetic defect at the pyridoxal phosphate coenzyme binding site of glutamic acid
decarboxylase resulting in brain GABA deficiency

Treatment

Minimum effective daily dose: varies between 10 and 100 mg orally
Convulsions cease within a few minutes when parenterally administered, within a few hours
when given orally
Single dose effective for a consistent period (2–5 days)
Interruption of treatment, seizures return
In case of (suspected) intrauterine convulsions, treatment of mother with pyridoxine is effective
(100 mg/day)

---

of GAD was assayed in kidney extract, and it is possible that the enzyme results reflect contributions from one, or both, molecular weight species of GAD. Therefore, demonstration of a renal GAD defect may not necessarily imply a brain enzyme defect.

GABA-transaminase deficiency

GABA-transaminase deficiency has been reported in two Flemish siblings (male and female) who had severe psychomotor retardation, hypotonia, hyperreflexia, seizures and accelerated linear growth (Jaeken et al 1984). Both died before 3 years of age. Autopsy findings revealed leukodystrophy in the brain of the male.

    Biochemical and enzymatic findings in GABA-transaminase deficiency are summarized in Table 3. Free GABA and β-alanine were elevated in plasma and CSF. Total GABA, unidentified GABA conjugates and homocarnosine were elevated in CSF. Elevated fasting plasma growth hormone (8–38 ng/ml, normal < 5) was consistent with the growth hormone-releasing effect of GABA, and was the likely cause of the increased linear growth in the two patients. GABA transaminase deficiency was documented in biopsied liver and leukocytes derived from the female (Table 3). The

Table 3   Biochemical and enzymatic findings in GABA-transaminase deficiency

Metabolites

Fluid	Metabolites	Elevation compared to control
CSF	Total GABA	3 ×
	Homocarnosine and 'other' GABA conjugates	3 ×
	β-Alanine	10 ×
	Free GABA	60 ×
Plasma	β-Alanine	4 ×
	Free GABA	9 ×

GABA-transaminase activity

Tissue	Subject	Residual GABA-transaminase activity
Biopsied liver	Patient	17%
White cells	Patient	3%
	Parents/sibling	13–37%

Abbreviations employed: CSF = cerebrospinal fluid; GABA = 4-aminobutyric acid

parents and an unaffected sibling demonstrated intermediate levels of GABA transaminase activity in isolated and cultured leukocytes, consistent with autosomal recessive inheritance. GABA transaminase in leukocytes, kidney and brain have similar K_m values for substrates GABA and 2-ketoglutaric acid (Gibson et al 1985), so demonstration of deficiency in other tissues may suggest a defect in brain GABA transaminase. A diagnosis of GABA transaminase deficiency cannot be accomplished in cultured skin fibroblasts or amniocytes because these cells do not express the enzyme. GABA transaminase is present, however, in chorionic villus tissue (Sweetman et al 1986). Molecular analysis of GABA transaminase deficiency has not been reported. However, a cDNA for GABA transaminase has been isolated from porcine brain, suggesting that a human brain cDNA should be isolated shortly (Kwon et al 1992).

Succinic semialdehyde dehydrogenase deficiency (4-hydroxybutyric aciduria)

Jakobs and coworkers (1981) identified the first case of 4-hydroxybutyric aciduria. To date, 32 cases have been identified, and the clinical, metabolic and enzymatic features for most patients are summarized in Table 4. In succinic semialdehyde dehydrogenase deficiency the oxidative step is blocked, and accumulated succinic semialdehyde is reduced to 4-hydroxybutyric acid (γ-hydroxybutyric acid) in a reaction catalysed by 4-hydroxybutyrate dehydrogenase (EC 1.1.1.61). Succinic semialdehyde dehydrogenase deficiency is an interesting inborn error of GABA metabolism because the biochemical hallmark, 4-hydroxybutyric acid, is a compound with unique neuropharmacological properties that have been summarized (Mamelak 1989). It is noteworthy that 4-hydroxybutyric acid, much like GABA, has also been found in the non-neural tissues of animals (Mamelak 1989), but its role there is unclear.

Table 4 Clinical, biochemical and enzymatic findings in succinic semialdehyde dehydrogenase deficiency

Clinical presentation[a]

Psychomotor retardation	25	Oculomotor apraxia	4
Delayed speech development	21	Macrocephaly	3
Hypotonia	16	Myopathy with ragged red fibres	2
Ataxia	16	Choreoathetosis	2
Hyporreflexia	12	Nystagmus	2
Convulsions	8	Conjunctival telangiectasias	2
Aggressive behaviour	6	Globus pallidus abnormalities	2
Hyperkinesis	5	(bilateral)	

[a]Values presented are the number of patients with that manifestation from a total of 25 from which clinical details were available. For the two patients with myopathy, only these two patients were biopsied.

Metabolite findings[b]

4-Hydroxybutyric acid	29	Glycolic acid	5
3,4-Dihydroxybutyric acid	10	3-Hydroxypropionic acid	4
Glycine (urine, plasma, CSF)	8	Glutaric acid	4
4,5-Dihydroxyhexanoic acid		Adipic acid	4
(various forms)	7	Suberic acid	4
3-Oxo-4-hydroxybutyric acid	5	Succinic semialdehyde	2
2,4-Dihydroxybutyric acid	5	Homovanillic acid	2

[b]Values presented are the number of patients with that metabolite from a total of 31 patients from which physiological fluids were obtained. CSF = cerebrospinal fluid.

Succinic semialdehyde dehydrogenase activity[c]

Cell type	Number of patients	Range of residual activity
Lymphocyte	16	0–19% (mean 4%)
Lymphoblast	21	0–6% (mean 2%)
Intact lymphoblast	14	4–12% (mean 8%)

[c]Enzyme results from a total of 21 patients studied.

The clinical presentation of succinic semialdehyde dehydrogenase deficiency includes delayed intellectual, motor, speech and language development (Table 4). Many patients manifest hypotonia and ataxia, but these and other clinical manifestations are variable. Ataxia, when present, may resolve with age (Hodson et al 1990). The key feature of succinic semialdehyde dehydrogenase deficiency is accumulation of 4-hydroxybutyric acid in body fluids, including urine, plasma and CSF, but quantitation of this metabolite may be hampered by loss due to volatility. In the patients analysed thus far, there is a tendency toward age-dependent urinary excretion of 4-hydroxybutyric acid, with high concentrations in younger patients and lower concentrations in older patients, although this may be a reflection of the change in ratio between brain and body mass with age (Gibson et al 1990; Jakobs et al 1990). It is tempting to correlate age-dependent 4-hydroxybutyric acid excretion with the observation that some younger succinic semialdehyde dehydrogenase-deficient patients are hypoactive, while some older patients are hyperactive, and even aggressive. This raises the possibility that 4-hydroxybutyric acid may bind to inhibitory sites in brain at high concentration and excitatory sites at low concentration.

Other metabolites indicative of the β- and α-oxidation of 4-hydroxybutyric acid may be variably detected in the urine of succinic semialdehyde dehydrogenase-deficient patients (Table 4, Figure 2). Dicarboxylic acids and 3-hydroxypropionic acid may be a source of diagnostic confusion by suggesting an abnormality of acyl-CoA or propionyl-CoA metabolism, but identification of 4-hydroxybutyric acid should establish the correct diagnosis. The explanation for glycinuria in some succinic semialdehyde dehydrogenase-deficient patients remains unclear. Glycolic acid in the urine of these patients is a product of the β-oxidation of 4-hydroxybutyric acid, consistent with the finding that rat liver mitochondria produce glycolyl-CoA when incubated with 4-hydroxybutyric acid (Vamecq et al 1990). Glycolic acid can be converted to glycine, and may be responsible for some increase of glycine in physiological fluids from succinic semialdehyde dehydrogenase-deficient patients.

Threo- and *erythro*-4,5-dihydroxyhexanoic acids (and the corresponding lactone forms) have been identified in the urine of succinic semialdehyde dehydrogenase-deficient patients (Brown et al 1987) (Table 4). This is of interest because 4,5-dihydroxyhexanoic acid (and its related forms) have not been identified in other heritable disorders of human metabolism, and these organic acids may be additional specific markers for succinic semialdehyde dehydrogenase deficiency, in addition to 4-hydroxybutyric acid. Brown and colleagues speculate that the dihydroxyhexanoic acid derivatives arise from condensation of a two-carbon moiety of pyruvate

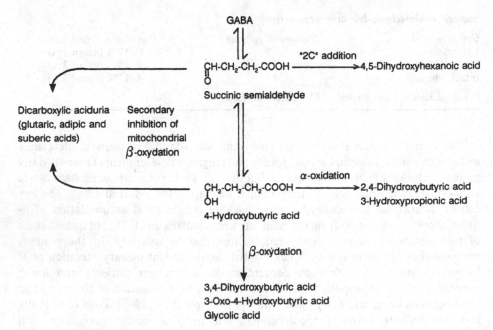

Figure 2 Probable origins of metabolites in succinic semialdehyde dehydrogenase deficiency

metabolism with accumulated succinic semialdehyde (Figure 2). The identification of metabolites in the urine of succinic semialdehyde dehydrogenase-deficient patients related to pathways of fatty-acid, pyruvate, and glycine metabolism suggests that the deficiency has metabolic consequences beyond the pathway of GABA metabolism.

Succinic semialdehyde dehydrogenase deficiency is demonstrated in isolated and cultured leukocytes (Table 4). Intermediate levels of enzyme activity for parents and family members of affected probands is consistent with autosomal recessive inheritance (Pattarelli et al 1988). Consanguinity of 7 in 21 families with affected probands, and sex distribution for deficient patients (11 male:16 female) are additional evidence of autosomal recessive inheritance. The prenatal diagnosis of an affected fetus with succinic semialdehyde dehydrogenase deficiency has been reported (Jakobs et al, 1993). 4-Hydroxybutyric acid was elevated in amniotic fluid, and succinic semialdehyde dehydrogenase activity was absent from cultured amniocytes and autopsied fetal brain, liver and kidney (Chambliss et al 1993). The irreversible GABA-transaminase inhibitor γ-vinylGABA (vigabatrin) has shown promise as a therapy for succinic semialdehyde dehydrogenase-deficient patients, with improvement of cerebellar signs in 5 out of 6 patients receiving the drug (Jakobs et al 1992; Howells et al 1992). However, long-term administration of vigabatrin should be monitored closely, because one of the metabolic consequences of administering this drug is increased GABA levels in the central nervous system.

Succinic semialdehyde dehydrogenase deficiency can manifest a mild to severe neurological course. Some patients have died, despite the fact that there is no episodic metabolic decompensation or metabolic acidosis. Despite the phenotypic heterogeneity observed, all patients demonstrate comparable low levels of residual activity in isolated or cultured leukocytes, suggesting that insight into phenotypic heterogeneity may be gained through mutation analysis. Such studies await the isolation and characterization of mammalian cDNAs encoding succinic semialdehyde dehydrogenase, although cDNAs encoding the enzyme have been isolated from *Escherichia coli* (Marek and Henson 1988; Bartsch et al 1990).

Homocarnosinosis

Homocarnosine is a brain-specific dipeptide of L-histidine and GABA whose physiological role is unknown (Skaper et al 1973). Homocarnosinosis was identified in a Norwegian family, including the mother and three affected siblings from a total of four children (Sjaastad et al 1976). Clinical manifestations in the affected siblings included spastic paraplegia, progressive mental deterioration and retinal pigmentation; the mother had no neurological findings. The disorder was characterized by homocarnosine accumulation in the CSF from the three patients and their mother (CSF homocarnosine concentration 50–75 mmol/L, normal < 3 mmol/L) (Sjaastad et al 1977). CSF carnosine was not increased in the patients, but urinary excretion of carnosine was elevated (Lunde et al 1982).

Homocarnosinase activity has not been convincingly demonstrated in normal human tissues (Lenney et al 1983). Serum carnosinase deficiency has been documented in patients without accumulation of homocarnosine, and with variable neurological findings (Murphey et al 1973; Cohen et al 1985). Carnosinase deficiency was, however,

712   *Jakobs et al.*

demonstrated in CSF from the three homocarnosinosis patients (Lenney et al 1983). Although homocarnosinosis resembles serum carnosinase deficiency, homocarnosinosis would intuitively appear to be a different disorder because the three affected Norwegian patients did not accumulate carnosine in the CSF. Detailed biochemical studies in additional patients with homocarnosinosis may clarify the underlying biochemistry.

Cerebrospinal fluid and the identification of inherited defects of GABA metabolism

With the exception of succinic semialdehyde dehydrogenase deficiency, many of the inherited disorders of GABA metabolism are unlikely to be diagnosed without accurate investigation of CSF metabolites. Besides free GABA, several other forms of GABA occur in human CSF: GABA-peptides (mainly homocarnosine, GABA-lysine and GABA-cystathionine), N-carboxyethylGABA and 2-pyrrolidinone. Of these, homocarnosine and 2-pyrrolidinone are quantitatively the most important. 2-Pyrrolidinone, the cyclization product of GABA, accounts for essentially all of the so-called 'unidentified conjugated GABA', i.e. the difference between total GABA and the sum of homocarnosine and free GABA. CSF GABA concentrations probably reflect brain GABA concentrations and are probably only minimally affected by alterations of the peripheral GABA concentration. Most methods for determining free GABA in brain tissue are insufficiently sensitive or specific for measuring the low free GABA concentrations in CSF. Two methods seem to have the required specificity and sensitivity: ion-exchange or reversed-phase chromatography with fluorescence detection (Carchon et al 1991) and (stable isotope dilution) gas chromatography–mass spectrometry (Kok et al 1993). Furthermore, the determination of free GABA in CSF is difficult because: (1) fractional collection of CSF specimens demonstrates a gradient effect of GABA concentration, with low concentration in the first and high concentration in subsequent specimens, which limits valid comparison between patient and controls; and (2) free GABA concentration in CSF shows artefactual increases due to enzymatic hydrolysis of homocarnosine unless specimens are deep-frozen immediately after withdrawal (Carchon et al 1991). At present it may be necessary to employ total GABA concentrations in CSF as a marker for alterations in brain GABA metabolism, but the possibility exists that patients with GABA-transaminase deficiency will be overlooked (Table 3).

Literature data on lumbar CSF free GABA concentrations in control children are summarized in Carchon et al (1991). Notably, children under 2 years of age have lower values than older children. The lower limit of lumbar CSF free GABA concentrations increases linearly from about 15 nmol/L in the neonatal period to about 65 nmol/L at the age of 1 year, and remains stable from there on. A similar evolution during the first year of life in CSF homocarnosine levels (an increase from 2.5 μmol/L to 6 μmol/L) and CSF total GABA (an increase from 4 μmol/L to 7 μmol/L) has been reported (Jaeken et al 1990).

CONCLUSIONS AND FUTURE DIRECTIONS

The study of patients with inherited disorders of GABA metabolism will provide new insight into biochemical and molecular aspects of GABA metabolism. Of the four diseases, the primary defect has been described convincingly only for succinic semialdehyde dehydrogenase and GABA-transaminase deficiencies, although documentation of deficient brain GABA-transaminase activity has yet to be presented. The underlying biochemistry of pyridoxine-dependent seizures and GAD deficiency, as well as homocarnosinosis, requires clarification. The multiple molecular forms of GAD pose a challenge to understanding which GAD protein is, or whether both are, involved in pyridoxine-dependent seizures. The investigation of molecular aspects of succinic semialdehyde dehydrogenase and GABA-transaminase deficiencies should be accomplished when human cDNAs encoding these proteins are isolated and characterized.

The identification of future patients with certain defects of GABA metabolism will depend upon the availability of accurate methods for determining free GABA concentrations in CSF. We suggest that more general paediatricians and neurometabolic specialists should recommend free and total GABA determination in CSF from patients with unexplained neurological abnormalities.

REFERENCES

Baekkeskov S, Aanstoot H-J, Christgau S et al (1990) Identification of the 64K autoantigen in insulin-dependent diabetes as the GABA-synthesizing enzyme glutamic acid decarboxylase. *Nature* 347: 151–156.

Bartsch K, von Johnn-Marteville A, Schulz A (1990) Molecular analysis of two genes of the *Escherichia coli gab* cluster: nucleotide sequence of the glutamate: succinic semialdehyde transaminase gene (*gabT*) and characterization of the succinic semialdehyde dehydrogenase gene (*gabD*). *J Bacteriol* 172: 7035–7042.

Brown GK, Cromby CH, Manning NJ, Pollitt RJ (1987) Urinary organic acids in succinic semialdehyde dehydrogenase deficiency: evidence of α-oxidation of 4-hydroxybutyric acid, interaction of succinic semialdehyde with pyruvate dehydrogenase and possible secondary inhibition of mitochondrial β-oxidation. *J Inher Metab Dis* 10: 367–375.

Bu D-F, Erlander MG, Hitz BC et al (1992) Two human glutamate decarboxylases, 65-kDa GAD and 67-kDa GAD, are each encoded by a single gene. *Proc Natl Acad Sci USA* 89: 2115–2119.

Carchon IIA, Jaeken J, Jansen E, Eggermont E (1991) Reference values for free gamma-aminobutyric acid determined by ion-exchange chromatography and fluorescence detection in the cerebrospinal fluid of children. *Clin Chim Acta* 201: 83–88.

Chambliss KL, Lee CF, Ogier H, Rabier D, Jakobs C, Gibson KM (1993) Enzymatic and immunologic demonstration of normal and defective succinic semialdehyde dehydrogenase activity in fetal brain, liver and kidney. *J Inher Metab Dis* 16: 523–526.

Cohen M, Hartlage PL, Krawiecki N, Roesel RA, Carter AL, Hommes FA (1985) Serum carnosinase deficiency: a non-disabling phenotype? *J Ment Defic Res* 29: 383–389.

Coker SB (1992) Postneonatal vitamin B6-dependent epilepsy. *Pediatrics* 90: 221–223.

Erdo SL (1985) Peripheral GABAergic mechanisms. *Trends Pharmacol Sci* 6: 205–208.

Gibson KM, Sweetman L, Nyhan WL, Jansen I, Jaeken J (1985) Demonstration of 4-aminobutyric acid aminotransferase deficiency in lymphocytes and lymphoblasts. *J Inher Metab Dis* 8: 204–208.

Gibson KM, Nyhan WL, Jaeken J (1986) Inborn errors of GABA metabolism. *BioEssays* 4: 24–27.

Gibson KM, Aramaki S, Sweetman L et al (1990) Stable isotope dilution analysis of 4-hydroxybutyric acid: an accurate method for quantification in physiological fluids and the prenatal diagnosis of 4-hydroxybutyric aciduria. *Biomed Environ Mass Spectrom* 19: 89–93.

Hodson AK, Gibson KM, Jakobs C (1990) Developmental resolution of ataxia in succinic semialdehyde dehydrogenase deficiency. *Ann Neurol* **28**: 438.

Howells D, Jakobs C, Kok RM, Wrennall J, Thompson GN (1992) Vigabatrin therapy in succinic semialdehyde dehydrogenase deficiency. *Mol Neuropharmacol* **2**: 181–184.

Iversen LL (1982) Neurotransmitters and CNS disease. *Lancet* **2**: 914–918.

Jaeken J (1990) Disorders of neurotransmitters. In Fernandes J, Saudubray J-M, Tada K, eds. *Inborn Metabolic Diseases: Diagnosis and Treatment*. New York: Springer-Verlag, 637–648.

Jaeken J, Casaer P, de Cock P et al (1984) Gamma-aminobutyric acid-transaminase deficiency: a newly recognized inborn error of neurotransmitter metabolism. *Neuropediatrics* **15**: 165–169.

Jaeken J, Casaer P, Haegele KD, Schechter PJ (1990) Review: Normal and abnormal central nervous system GABA metabolism in childhood. *J Inher Metab Dis* **13**: 793–801.

Jakobs C, Bojasch M, Moench E, Rating D, Siemes H, Hanefeld F (1981) Urinary excretion of gamma-hydroxybutyric acid in a patient with neurological abnormalities: the probability of a new inborn error of metabolism. *Clin Chim Acta* **111**: 169–178.

Jakobs C, Smit LME, Kneer J, Michael T, Gibson KM (1990) The first adult case with 4-hydroxybutyric aciduria. *J Inher Metab Dis* **13**: 341–344.

Jakobs C, Michael T, Jaeger E, Jaeken J, Gibson KM (1992) Further evaluation of Vigabatrin therapy in 4-hydroxybutyric aciduria. *Eur J Pediatr* **151**: 466–468.

Jakobs C, Ogier H, Rabier D, Gibson KM (1993) Prenatal detection of succinic semialdehyde dehydrogenase deficiency (4-hydroxybutyric aciduria). *Prenat Diagn* **13**: 150.

Johnston MV, Singer HS (1982) Brain neurotransmitters and neuromodulators in pediatrics. *Pediatrics* **70**: 57–68.

Kok RM, Howells DW, Heuvel CCM vd, Guérand WS, Thompson GN, Jakobs C (1993) Stable isotope dilution analysis of GABA in CSF using simple solvent extraction and electron capture negative ion mass fragmentography. *J Inher Metab Dis* **16**: 508–512.

Kwon O-S, Park J, Churchich JE (1992) Brain 4-aminobutyrate aminotransferase. *J Biol Chem* **267**: 7215–7216.

Lenney JF, Peppers SC, Kucera CM, Sjaastad O (1983) Homocarnosinosis: lack of serum carnosinase is the defect probably responsible for elevated brain and CSF homocarnosine. *Clin Chim Acta* **132**: 157–165.

Lunde H, Sjaastad O, Gjessing L (1982) Homocarnosinosis: hypercarnosinuria. *J Neurochem* **38**: 242–245.

Mamelak M (1989) Gammahydroxybutyrate: an endogenous regulator of energy metabolism. *Neurosci Biobehav Rev* **13**: 187–198.

Marek LE, Henson JM (1988) Cloning and expression of the *Escherichia coli* K-12 *GAD* gene. *J Bacteriol* **170**: 991–994.

Murphey WH, Lindmark DG, Patchen LI, Housler ME, Harrod EK, Mosovich L (1973) Serum carnosinase deficiency concomitant with mental retardation. *Pediatr Res* **7**: 601–606.

Pattarelli PP, Nyhan WL, Gibson KM (1988) Oxidation of $[U^{14}C]$succinic semialdehyde in cultured human lymphoblasts: measurement of residual succinic semialdehyde dehydrogenase activity in 11 patients with 4-hydroxybutyric aciduria. *Pediatr Res* **24**: 455–460.

Scriver CR, Perry TL (1989) Disorders of ω-amino acids in free and peptide-linked forms. In Scriver CR, Beaudet AL, Sly WS, Valle D, eds. *The Metabolic Basis of Inherited Disease*, 6th edn. New York: McGraw-Hill, 755–771.

Sjaastad O, Berstad J, Gjesdahl P, Gjessing L (1976) Homocarnosinosis. 2. A familial metabolic disorder associated with spastic paraplegia, progressive mental deficiency, and retinal pigmentation. *Acta Neurol Scand* **53**: 275–290.

Sjaastad O, Gjessing L, Berstad JR, Gjesdahl P (1977) Homocarnosinosis. 3. Spinal fluid amino acids in familial spastic paraplegia. *Acta Neurol Scand* **55**: 158–162.

Skaper SD, Das S, Marshall FD (1973) Some properties of a homocarnosine-carnosine synthetase isolated from rat brain. *J Neurochem* **21**: 1429–1445.

Sweetman FR, Gibson KM, Sweetman L et al (1986) Activity of biotin-dependent and GABA metabolizing enzymes in chorionic villus samples: potential for first trimester prenatal diagnosis. *Prenat Diagn* **6**: 187–194.

Tanaka R, Okumura M, Arima J, Yamakura S, Momoi T (1992) Pyridoxine-dependent seizures: report of a case with atypical clinical features and abnormal MRI scans. *J Child Neurol* 7: 24–28.

Vamecq J, Draye J-P, Poupaert JH (1990) Studies on the metabolism of glycolyl-CoA. *Biochem Cell Biol* 68: 846–851.

Yoshida T, Tada K, Arakawa T (1971) Vitamin B$_6$ dependency of glutamic acid decarboxylase in the kidney from a patient with vitamin B$_6$-dependent convulsion. *Tohoku J Exp Med* 104: 195–198.

J. Inher. Metab. Dis. 16 (1993) 716–723
© SSIEM and Kluwer Academic Publishers. Printed in the Netherlands

Excitotoxicity, Energy Metabolism and Neurodegeneration

A. C. LUDOLPH[1], M. RIEPE[2] and K. ULLRICH[3]

[1]*Department of Epileptology, University of Bonn, Sigmund Freud-Strasse 25,
53 Bonn, Germany;* [2]*Wadsworth Center, Albany, New York, USA;* [3]*Department of
Pediatrics, University of Münster, Germany*

Summary: There is increasing evidence that the neurotoxic effects of excitatory
amino acids and their analogues are part of the pathogenesis of neuronal
degeneration in acute and chronic neurological disease. Recent studies indicate
that activation of excitatory amino acid receptors is also induced in the
mechanism of neuronal damage induced by impairment of cellular energy
metabolism. This article briefly summarizes the evidence for the presence of
such a mechanism and discusses metabolic diseases in which excitatory amino
acids alone or in combination with energy deficiency could play a pathogenetic
role. In these and other metabolic diseases, antagonists to excitatory amino
acid receptors may offer a therapeutic opportunity; however, there are potential
limits that may prevent chronic use.

EXCITOTOXICITY

The concept of *excitotoxicity* describes the potential cell damage induced by
excitatory amino acid neurotransmitters like glutamate. It means that this excitant
neurotransmitter not only has a key role in plasticity during learning and development
but also has the potential to injure and destroy neurons (Choi and Rothman 1990;
Jahr and Lester 1992; McDonald and Johnston 1990; Meldrum and Garthwaite
1990; Olney 1980; Rothman 1992). Olney (1980) showed that intracerebral injection
of excitatory amino acids acutely induces lesions characterized by massive swelling
and vacuolation of the cytoplasm, mitochondrial expansion, dendritic ballooning and
focal clumping of nuclear chromatin. Axons, presynaptic terminals and non-neuronal
cells were spared. The development of selective antagonists to excitatory amino acid
receptors provided the tools to show that the neurotoxic effects are receptor-mediated
and a consequence of excessive receptor activation (Meldrum and Garthwaite 1990).

Toxic and other effects of glutamate are mediated through pharmacologically
distinct receptors. Synaptic activation of the *N*-methyl-D-aspartate (NMDA) receptor
ion channel complex is characterized by a slow rise time, whereas the non-NMDA
receptors activated by AMPA (α-amino-3-hydroxy-5-methyl-4-isoxazolepropionate)
and kainate mediate fast transmission through an ionotropic channel. Activation of
the metabotropic receptor results in signal transduction via a GTP-binding protein;
the relation of this receptor type to neurotoxic effects is currently being investigated
(Koh et al 1991a,b; Schoepp et al 1991). Progress in the molecular definition of

different glutamate receptor types (Barnes and Henley 1992) shows that within these groupings multiple subclasses with differing anatomical, physiological and pharmacological profiles can be identified. A detailed description is beyond the scope of this article (see Barnes and Henley 1992).

In brief, the concept of *acute excitotoxicity* postulates that in an early phase of cell activation sodium enters the cell. This is followed by passive influx of chloride and water, which in turn produces osmotic swelling (Rothman 1992). This initial phase is partly reversible. In a second phase the intracellular concentration of free calcium increases due to calcium influx and release from intracellular stores. This increase seems to be a crucial step, although the subsequent events that lead to cell death are unknown. Activation of Ca^{2+}-dependent proteases may be an important mechanism (Lee et al 1991; Manev et al 1991). There is evidence that the *acute* degenerative changes following status epilepticus, hypoglycaemia and cerebral ischaemia may partly result from endogenously released excitatory amino acids activating NMDA and non-NMDA receptors (Choi and Rothman 1990; Kaku et al 1991; Meldrum and Garthwaite 1990; Mosinger et al 1991). The observations that the acute cytopathology of these lesions is similar to that induced by excitatory amino acids and that antagonists to excitatory amino acids ameliorate neuronal damage support the proposal that excitatory amino acids play a role in the pathogenesis.

An observation also supporting a role of excitotoxicity in acute nerve cell degeneration is the outbreak of domoic acid poisoning that occurred in Canada in 1987 (Teitelbaum et al 1990). Domoic acid is a structural analogue of the non-NMDA agonist kainic acid. The effects of intoxication in individuals who consumed domoic acid-contaminated mussels were characterized by an acute encephalopathy with seizures and myoclonus. Some patients developed a chronic neurological deficit with amnesia. Four subjects died from domoic acid poisoning and cortical pathology correlated with the location of greatest density of kainate receptors. The clinical syndrome and pathological findings may be explained by a direct activating effect of postsynaptic kainate receptors; elevated extracellular levels of glutamate secondary to a presynaptic effect of domoate may also contribute.

The relation between compounds with acute excitatory effects and *chronic* diseases is less convincing. The cause of the upper motor neuron disease lathyrism is the chronic consumption of a plant that contains large amounts of excitatory amino acid — β-oxalylamino-L-alanine (BOAA) — an AMPA agonist (Ludolph et al 1987). The precise pathogenetic relation between acute *in vitro* and *in vivo* effects of this excitatory amino acid and the stereotyped irreversible but non-progressive clinical picture developing after 2–3 months of chronic consumption is unclear. The neurochemical and pathological similarities of acute excitotoxic lesions in the basal ganglia to the pattern of vulnerability in Huntington's chorea provides a tool to explore the pathogenesis of the disease; however, all suspected compounds can only hypothetically explain a chronic effect in humans (Beal 1992).

EXCITOTOXICITY AND ENERGY DEFICIENCY

Recent evidence supports the view that neuronal damage induced by cellular energy depletion develops via a mechanism that includes activation of glutamate receptors

(Beal 1992; Henneberry et al 1989). The *in vitro* work of Henneberry's group suggested that cellular energy deficits lead to an activation of the NMDA channel by a reduction of the resting membrane potential: this induces a relief of the voltage-dependent magnesium block of the channel and allows ions to enter the cell persistently (Henneberry et al 1989). This hypothesis is supported by recent experimental results that show that the mitochondrial toxins MPP^+ (Turski et al 1991), 3-nitropropionic acid (Ludolph et al 1992), aminooxyacetic acid (Beal et al 1991) and cyanide (Zeevalk and Nicklas 1991, 1992) induce morphological changes comparable to that associated with excitotoxic damage. In each case, damage could be at least partly prevented by antagonists at glutamate receptors. In the chick retinal preparation, Zeevalk and Nicklas induced graded metabolic stress by blockade of electron transport by potassium cyanide (Zeevalk and Nicklas 1991, 1992). Mild metabolic stress induced histopathology similar to that seen after glutamate agonist treatment and could be completely prevented by NMDA antagonists or by elevating extracellular Mg^{2+}. Extracellular glutamate did not increase, indicating that the opening of the magnesium block of the NMDA channel is sufficient to explain the observed effects. Severe compromise of energy metabolism causing more extensive neuronal damage was accompanied by elevations of extracellular glutamate, and was only partially attenuated by antagonists to excitatory amino acid receptors.

Electrophysiological studies after inhibition of oxidative phosphorylation by 3-nitropropionic acid showed that after a reduction of cellular energy a temporary hyperpolarization due to activation of ATP-sensitive potassium channels is followed by depolarization of the cell membrane (Riepe et al 1992). Responses to extracellularly applied glutamate are decreased (Riepe et al 1992, and unpublished). Since ion pumps are critically dependent on energy supply, this observation indicates that an increase of intracellular ion concentration as a result of increased influx relative to efflux, and not the activation of glutamate receptors alone, plays a major role in the degenerative process. Antagonists at ionotropic glutamate receptors applied *late* after onset of the terminal depolarization prolong further depolarization but fail to reverse the membrane potential. However, if an antagonist at the NMDA receptor, like MK-801, is applied during *early* depolarization, neurons temporarily repolarize and the terminal depolarization under conditions of *ongoing* inhibition of energy metabolism is prolonged (Riepe et al 1992, and unpublished).

In summary, it seems that antagonists to excitatory amino acid receptors can partly prevent neuronal damage induced by energy deficits. The effect is apparently restricted to early or mild forms of damage; no major influence is expected in late stages of nerve cell depolarization.

EXCITOTOXICITY AND METABOLIC DISEASES

Neurodegeneration as a consequence of an activation of NMDA and non-NMDA receptors is a possible pathogenetic mechanism in some metabolic diseases. There is now definite evidence that activation of excitatory amino acid receptors plays a role in the pathophysiology of non-ketotic hyperglycinaemia.

The NMDA receptor subtype is associated with a cationic ion channel that is permeable to calcium (Figure 1). For channel opening the binding of synaptically

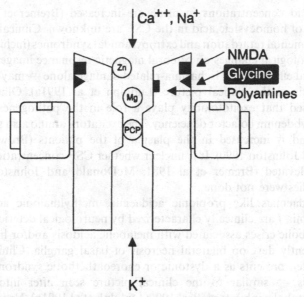

Figure 1 Simplified schema (modified according to Barnes and Henley 1992) of the role of the NMDA receptor in the pathogenesis of non-ketotic hyperglycinaemia (NKH). Recent results indicate that activation of the strychnine-insensitive glycine recognition site is part of the pathogenesis of seizures in NKH. Seizures can be treated by blockers of the ion channel of the NMDA receptor complex ('PCP-like antagonists')

released glutamate and two glycine molecules is required. Glycine binds to a recognition site that is not sensitive to strychnine. The bulk of experimental evidence indicates that extracellular glycine is tonically present *in vivo* and therefore it was uncertain whether even a more than 10-fold increase of glycine — as observed in the cerebrospinal fluid (CSF) of patients with non-ketotic hyperglycinaemia (NKH) — has an effect on NMDA receptor activation. Recent reports on therapeutic attempts in single patients show that the seizures associated with the disease can be effectively treated with a non-competitive glutamate antagonist and ion channel blocker like dextromorphane (Schmitt et al 1992) or ketamine (Ohta et al 1990). Such an effect could not be shown in treatment attempts with strychnine. The partial or minor success with benzodiazepine treatment (Matalon et al 1983) does not necessarily contradict the conclusion that the development of seizures in NKH is mediated by activation of the NMDA receptor since benzodiazepines may interfere with presynaptic glutamate release. We do not know whether neuronal degeneration in NKH is also closely related to glutamate receptor activation and can be treated by an optimized future drug regimen. A systematic therapeutic trial is clearly necessary.

L-Homocysteic acid is a potent glutamate analogue and candidate neurotransmitter (Olney 1980). This amino acid is taken up by a specific low-affinity system (Cox et al 1977) and antagonists to the NMDA receptor subtype greatly reduce responses induced by this excitatory compound (Knopfel et al 1987; Watkins and Evans 1981). In patients with homocystinuria due to cystathionine β-synthase deficiency, homocysteic acid concentrations in urine are increased (Bremer et al 1981), but concentrations of homocysteic acid in the CSF are unknown. Clinical features such as convulsions, mental retardation and extrapyramidal syndromes (including dystonia) without morphological changes on cerebral magnetic resonance images indicate that pharmacological effects — not ischaemia-related damage alone — may partly explain the pathogenesis of the clinical picture (Ludolph et al 1991a). Olney (1980) had already proposed that excitotoxicity plays a role in the pathogenesis of cerebral damage in molybdenum cofactor deficiency. The excitatory amino acid sulphocysteine accumulates and is increased in the plasma of the patients (Brown et al 1989; McDonald and Johnston 1990). It is unclear whether CSF concentrations of cysteine sulphate are elevated (Bremer et al 1981; McDonald and Johnston 1990), and systematic studies were not done.

Organic acidaemias like propionic acidaemia, methylmalonic acidaemia, and glutaric acidaemia I are clinically characterized by neurological deterioration following acute metabolic crises associated with metabolic acidosis and/or hypoglycaemia. Patients frequently develop bilateral necrosis of basal ganglia. Clinically, glutaric acidaemia I often presents as a dystonic or choreoathethotic syndrome following a metabolic crisis — similar to the clinical picture seen after intoxication with endogenous metabolic inhibitors (Beal 1992; Ludolph et al 1991b). Neuropathological findings characteristically include a spongy myelinopathy and bilaterally symmetric striatal degeneration. In a recent report on an autopsy, morphological alterations consistent with the presence of an excitotoxic mechanism were described (Amir et al 1992). The acute states of metabolic derangement are associated with potential acute or chronic accumulation of excitatory transmitters and their analogues.

In vitro, glutaric acid inhibits glutamate decarboxylase in concentrations found in the disease (Stokke et al 1976). In addition, glutaric acid inhibits glutamate uptake in the synaptosomal fraction (Balcar and Johnston 1972; Bennett et al 1973), which also increases local glutamate concentrations dose-dependently. Also the metabolic block in glutaric acidaemia type I induces an accumulation of 2-amino-3-carboxymuconic semialdehyde, which could lead to an enhanced synthesis of quinolinic acid, a potent glutamate analogue and neurotoxin both *in vitro* and *in vivo* (Heyes 1987). However, the quinolinic acid level was reportedly normal in the CSF of a single patient with glutaric acidaemia type I (Land et al 1992). In propionic and methyl malonic acidaemia the energy deficiency state may be aggravated by elevated glycine concentrations due to inhibition of glycine cleavage, glycine–serine interconversion or a negative nitrogen balance (Bremer et al 1981).

A THERAPEUTIC WINDOW?

Successful treatment of seizures associated with NKH is possible by application of glutamate antagonists (Schmitt et al 1992; Ohta et al 1990). It is uncertain whether control of seizures is also accompanied by a therapeutic influence on permanent neurological deficits such as mental retardation. Even if a dosage and compound could be found that positively influences these deficits, prenatal activation of excitotoxic mechanisms (Garcia-Castro et al 1982) might already have severe negative effects on development and could lead to irreversible neurological deficits. In addition, there is uncertainty about the severity of side-effects of treatment with glutamate antagonists in early critical developmental phases. Glutamate also has a major neurotrophic influence and has a role in the development of neuronal circuitries and synaptic plasticity (Jahr and Lester 1992; McDonald and Johnston 1990). Therefore, chronic use of glutamate antagonists has major risks that might outweigh their benefits.

Another concern is the possibility that non-competitive and competitive glutamate antagonists induce neuronal vacuolation in cingulate and retrosplenial cortex (Olney et al 1989; Sharp et al 1991). However, a recent study showed that cholinergic antagonists and increase of GABAergic inhibition can prevent this potential side-effect (Olney et al 1991). Therapeutic intervention by manipulation of levels of the endogenous glutamate antagonist kynurenic acid — another tryptophan metabolite — may also offer a future therapeutic opportunity (Nozaki and Beal 1992).

In states of energy deficiency, the bulk of evidence supports a view that antagonists to excitatory amino acid receptors have a well-defined protective effect. Experimentally, this effect is limited to mild electrophysiological, biochemical, and morphological changes resulting from energy deficiency (Riepe et al 1992, and unpublished; Zeevalk and Nicklas 1991, 1992). In severe states of neuronal energy deficits other mechanisms than activation of excitatory receptors alone seem to lead to cell damage. Therefore, only the consequences of temporary metabolic alterations should be successfully prevented by such a therapeutic schema.

In summary, we suggest that excitotoxic mechanisms play a role in the pathogenesis of some inherited metabolic diseases and emphasize that not only classical candidate diseases like NKH and homocystinuria should be considered. Recent evidence indicates that diseases in which energy deficiency plays a role in the pathogenesis are also related to an excitotoxic mechanism.

REFERENCES

Amir N, Elpeleg RN, Joseph A, Soffer D (1992) Glutaric aciduria type I: Correlation between neuroradiological features and pathological findings. *Abstracts of the 30th Annual Symposium of SSIEM*, P86.

Balcar VJ, Johnston GAR (1972) The structural specificity of the high affinity uptake of L-glutamate and L-aspartate by rat brain slices. *J Neurochem* **19**: 2657–2666.

Barnes JM, Henley JM (1992) Molecular characteristics of excitatory amino acid receptors. *Progr Neurobiol* **39**: 113–133.

Beal MF (1992) Does impairment of energy metabolism result in excitotoxic neuronal death in neurodegenerative illnesses? *Ann Neurol* **31**: 119–130.

Beal MF, Swartz KJ, Hyman BT, Storey E, Finn SF, Koroshetz W (1991) Aminooxyacetic acid results in excitotoxin lesions by a novel indirect mechanism. *J Neurochem* **57**: 1068–1073.

Bennett JP, Logan WJ, Snyder SH (1973) Amino acids as central nervous transmitters: The influence of ions, amino acid analogues, and ontogeny on transport systems for L-glutamic and aspartic acids and glycine into central nervous synaptosomes of the rat. *J Neurochem* **21**: 1533–1550.

Bremer HJ, Duran M, Kamerling JP, Przyrembel H, Wadman SK (1981) *Disturbances of Amino Acid Metabolism: Clinical Chemistry and Diagnosis*. Baltimore–Munich: Urban & Schwarzenberg.

Brown GK, Scholem RD, Croll HB, Wraith JE, McGill JJ (1989) Sulfite oxidase deficiency: clinical, neuroradiologic, and biochemical features in two new patients. *Neurology* **39**: 252–257.

Choi DW, Rothman SM (1990) The role of glutamate neurotoxicity in hypoxic-ischemic cell death. *Annu Rev Neurosci* **13**: 171–182.

Cox DWG, Headley MH, Watkins JC (1977) Actions of L- and D-homocysteate in rat CNS: a correlation between low-affinity uptake and the time courses of excitation by microelectrophoretically applied L-glutamate analogues. *J Neurochem* **29**: 579–588.

Garcia-Castro J-M, Isales-Forsythe CM, Levey HL et al (1982) Prenatal diagnosis of non-ketotic hyperglycinemia. *N Engl J Med* **306**: 79–81.

Henneberry RC, Novelli A, Cox JA, Lysko PG (1989) Neurotoxicity at the *N*-methyl-D-aspartate receptor in energy-compromised neurons: An hypothesis for cell death in aging and disease. *Ann NY Acad Sci* **568**: 225–233.

Heyes MP (1987) Hypothesis: A role for quinolinic acid in the neuropathology of glutaric aciduria type I. *Can J Neurol Sci* **14**: 441–443.

Jahr L, Lester RA (1992) Synaptic excitation mediated by glutamate-gated ion channels. *Curr Opin Neurobiol* **2**: 270–274.

Kaku DA, Goldberg MP, Choi DW (1992) Antagonism of non-NMDA receptors augments the neuroprotective effect of NMDA receptor blockade in cortical cultures subjected to prolonged deprivation of oxygen and glucose. *Brain Res* **554**: 344–347.

Knopfel T, Zeise ML, Cuenod M, Zieglgansberger W (1987) L-Homocysteic acid but not L-glutamate is an endogenous *N*-methyl-D-aspartic acid preferring agonist in rat neocortical neurons in vitro. *Neurosci Lett* **81**: 188–192.

Koh, J-Y, Palmer E, Lin A, Cotman CW (1991a) A metabotropic glutamate receptor agonist does not mediate neuronal degeneration in cortical culture. *Brain Res* **561**: 338–343.

Koh J-Y, Palmer E, Cotman CW (1991b) Activation of the metabotropic glutamate receptor attenuates *N*-methyl-D-aspartate neurotoxicity in cortical cultures. *Proc Natl Acad Sci USA* **88**: 9431–9435.

Land JM, Goulder P, Johnson AW, Hockaday J (1992) Quinolinate and glutaric aciduria type I. *Abstracts of the 30th Annual Symposium of SSIEM*, P88.

Lee KS, Frank S, Vanderklish P, Arai A, Lynch G (1991) Inhibition of proteolysis protects hippocampal neurons from ischemia. *Proc Natl Acad Sci USA* **88**: 7233–7237.

Ludolph AC, Hugeon J, Dwivedi MP, Schaumburg HH, Spencer PS (1987) Studies on the etiology and pathogenesis of motor neuron diseases. Lathyrism. Clinical findings in established cases. *Brain* **110**: 149–165.

Ludolph AC, Ullrich K, Bick U, Fahrendorf G, Przyrembel H (1991a) Functional and morphological deficits in late-treated patients with homocystinuria: a clinical, electrophysiological, and MRI study. *Acta Neurol Scand* **83**: 161–165.

Ludolph AC, He F, Spencer PS, Hammerstad J, Sabri M (1991b) 3-Nitropropionic acid – exogenous animal neurotoxin and possible human striatal toxin. *Can J Neurol Sci* **18**: 492–498.

Ludolph AC, Seelig M, Ludolph AG, Sabri MI, Spencer PS (1992) Cellular energy deficits and excitotoxic lesions induced by 3-nitropropionic acid in vitro. *Neurodegeneration* **1**: 155–161.

Manev H, Favaron M, Siman R, Guidotti A, Costa E (1991) Glutamate neurotoxicity is independent of calpain I inhibition in primary cultures of cerebellar granule cells. *J Neurochem* **57**: 1288–1295.

Matalon R, Naidu S, Hughes JR, Michals K (1983) Non-ketotic hyperglycinemia: Treatment with diazepam — a competitor for glycine receptors. *Pediatrics* **71**: 581–584.

McDonald JW, Johnston MV (1990) Physiological and pathophysiological roles of excitatory amino acids during central nervous system development. *Brain Res Rev* **15**: 41–70.

Meldrum B, Garthwaite J (1990) Excitatory amino acid neurotoxicity and neurodegenerative disease. *TIPS* **11**: 379–387.

Mosinger JL, Price MT, Bai HY, Xiao H, Wozniak DF, Olney JW (1991) Blockade of both NMDA and non-NMDA receptors is required for optimal protection against ischemic neuronal degeneration in the in vivo adult mammalian retina. *Exp Neurol* 113: 10–17.

Nozaki K, Beal MF (1992) Neuroprotective effects of L-kynurenine on hypoxia-ischemia and NMDA lesions in neonatal rats. *J Cereb Blood Flow Metab* 12: 400–407.

Ohta Y, Ochi N, Mizutani N, Hayakawa C, Watanabe K (1990) Non-ketotic hyperglycinemia: Therapeutic attempt with NMDA receptor antagonist. *Abstracts of the 5th International Congress on Inherited Metabolic Diseases*, P79.

Olney JW (1980) Excitotoxic mechanisms of neurotoxicity. In Spencer PS, Schaumburg HH, eds. *Clinical and Experimental Neurotoxicology*. Baltimore: Williams and Wilkins, 272–294.

Olney JW, Labruyere J, Price MT (1989) Pathological changes induced in cerebrocortical neurons by phencyclidine and related drugs. *Science* 244: 1360–1362.

Olney JW, Labruyere J, Wang G, Wozniak DF, Price MT, Sesma MA (1991) NMDA antagonist neurotoxicity: Mechanism and prevention. *Science* 254: 1515–1518.

Riepe M, Hori N, Ludolph AC, Carpenter DO, Spencer PS (1992) Inhibition of energy metabolism by 3-nitropropionic acid activates ATP-sensitive potassium channels. *Brain Res* 586: 61–66.

Rothman SM (1992) Excitotoxins: Possible mechanisms of action. *Ann NY Acad Sci* 648: 132–139.

Schmitt B, Steinmann B, Gitzelmann R, Thun-Hohenstein L, Dumermuth G (1992) Dextromethorphan, a N-methyl-D-aspartate antagonist, in the treatment of non-ketotic hyperglycinemia. *Abstracts of the 30th Annual Symposium of SSIEM*, O5.

Schoepp DD, Johnson BG, Salhoff CR, McDonald JW, Johnston MV (1991) In vitro and in vivo pharmacology of trans- and cis-(+)-1-amino-1,3-cyclopentanedicarboxylic acid: Dissociation of metabotropic and ionotropic excitatory amino acid receptor effects. *J Neurochem* 56: 1789–1796.

Sharp FR, Jasper P, Hall J, Noble L, Sagar SM (1991) MK-801 and ketamine induce heat shock protein HSP72 in injured neurons in posterior cingulate and retrosplenial cortex. *Ann Neurol* 30: 801–809.

Stokke O, Goodman SI, Moe PG (1976) Inhibition of brain glutamate decarboxylase by glutarate, glutaconate, and β-hydroxyglutarate: Explanation of the symptoms in glutaric aciduria. *Clin Chim Acta* 66: 411–415.

Teitelbaum JS, Zatorre RJ, Carpenter S et al (1990) Neurological sequelae of domoic acid intoxication due to the ingestion of contaminated mussels. *N Engl J Med* 322: 1781–1787.

Turski L, Bressler K, Rettig K-J, Löschmann P-A, Wachtel H (1991) Protection of substantia nigra from MPP^+ neurotoxicity by N-methyl-D-aspartate antagonists. *Nature* 349: 414–418.

Watkins JC, Evans RH (1981) Excitatory amino acid neurotransmitters. *Annu Rev Pharmacol Toxicol* 21: 165–204.

Zeevalk GD, Nicklas WJ (1991) Mechanisms underlying initiation of excitotoxicity associated with metabolic inhibition. *J Pharmacol Exp Ther* 257: 870–878.

Zeevalk GD, Nicklas WJ (1992) Evidence that the loss of the voltage-dependent Mg^{2+} block at the N-methyl-D-aspartate receptor underlies receptor activation during inhibition of neuronal metabolism. *J Neurochem* 59: 1211–1220.

J. Inher. Metab. Dis. 16 (1993) 724–732
© SSIEM and Kluwer Academic Publishers. Printed in the Netherlands

An Introduction to the Molecular Basis of Inherited Myelin Diseases

J.-M. Matthieu

Service de Pédiatrie, Centre Hospitalier Universitaire Vaudois, CH-1011 Lausanne, Switzerland

Summary: The myelin sheath is an extension of a plasma membrane tightly wrapped around axons. It facilitates conduction while conserving space and energy. Myelin is characterized by a high lipid content (80% of dry weight). Most myelin proteins are unique to that structure and some of them are restricted to the central or peripheral nervous system. In this review a few examples of inherited metabolic disorders affecting the oligodendrocyte and/or the Schwann cells are presented. Emphasis is placed on mutations in animals that represent invaluable models for investigating the molecular mechanisms of inherited myelin diseases in humans.

The purpose of this report is to present a few examples of inherited metabolic disorders affecting primarily the myelin in the central and peripheral nervous systems. Numerous metabolic diseases can secondarily induce glial cell loss due to the accumulation of a toxic metabolite or to the lack of a substrate or co-factor. These diseases (e.g. phenylketonuria) should not be considered as primary hereditary metabolic diseases of myelin.

Before presenting a few examples of primary myelin diseases expressed in animals and in humans, some basic knowledge of the structure, the function and the composition of the myelin sheath will be reviewed.

MORPHOLOGY, FUNCTION AND COMPOSITION OF MYELIN

The myelin sheath is a spiral structure tightly wrapped around the axon (Figure 1). It is a greatly extended and modified plasma membrane produced by a glial cell: the oligodendrocyte in the central nervous system (CNS) and the Schwann cell in the peripheral nervous system (PNS). The cytoplasm is restricted to the outer and lateral loops and, in the PNS, to additional tubes of cytoplasm that make up the Schmidt–Lantermann clefts.

On electron micrographs, a cross-section of a myelinated axon reveals the typical repeat structure with the alternation of minor and major dense lines. The minor dense line is formed by the apposition of the external surface of the plasma membrane. Owing to the lack of cytoplasm, the cytoplasmic sides of the membrane fuse and form the major dense line.

Myelin is an insulator in myelinated fibres. Depolarization of the axolemma (the

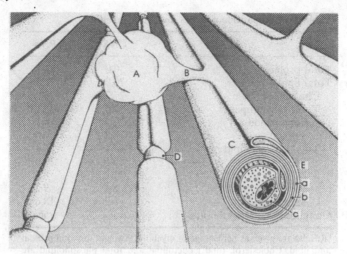

Figure 1 Schematic representation of myelinated axons in the central nervous system. A, oligodendrocyte cell body. B, cytoplasmic process maintaining the contact between a myelin sheath segment and its cell. C, external layer of the myelin sheath. D, Node of Ranvier, the only place for depolarization along a myelinated axon. E, cross-section through a myelinated axon wrapped by three compact myelin layers; (a) major dense line formed by the fusion of the cytoplasmic face of the membrane; (b) intraperiod line, apposition of the extracellular faces of adjacent lamellae; (c) axon containing neurotubules, neurofilaments and one mitochondrion

axon membrane) occurs between myelin segments. Therefore, action currents 'jump' from one interruption (Node of Ranvier) to the next. This form of impulse propagation is called saltatory conduction. Therefore, the wave of depolarization is much more rapid in myelinated fibres than in unmyelinated ones, where the entire axonal membrane is depolarized in a continuous sequential fashion. Furthermore, this localized depolarization is energy-sparing. Conduction velocity in myelinated fibres is approximately proportional to the diameter, whereas in unmyelinated fibres it is proportional to the square root of the diameter. Therefore, for high-speed conduction myelinated fibres allow space-saving. In brief, myelin facilitates conduction while conserving space and energy. This has been a decisive evolutionary advantage that allowed mammals to develop very sophisticated neural networks whilst keeping a reasonable brain size.

Myelin lipids

In contrast to the other plasma membranes, CNS and PNS myelin are characterized by a high lipid content (ca. 80% of dry weight) and only 20% of protein. This property has been used to isolate myelin by high-speed centrifugation of nervous tissue homogenates on continuous or discontinuous density gradients.

Table 1 gives the lipid composition of CNS and PNS myelin. If there are, strictly speaking, no myelin-specific lipids, cerebroside (galactosylceramide) is the most typical myelin lipid. During brain development its concentration is directly proportional to the amount of myelin deposited. The main difference between CNS and PNS myelin

Table 1   Composition of CNS and PNS myelin[a]

	CNS	PNS
Total protein	30.0	28.7
Total lipid	70.0	71.3
Cholesterol	27.7	23.0
Total galactolipid	27.5	22.1
Cerebroside	22.7	16.1
Sulphatide	3.8	6.0
Total phospholipid	43.1	54.9
PE	16.6	19.0
PC	11.2	8.1
PS + PI	6.4	9.2
Sphingomyelin	8.9	18.6
Gangliosides, paraglobosides	trace	trace

[a]Results are given as percentage of myelin dry weight for total protein and lipid. Cholesterol, total galactolipid and total phospholipid are expressed as percentage of total lipid. Individual galactolipids and phospholipids are also expressed as percentage of total lipids

resides in lower concentrations of galactolipids (cerebroside and sulphatide) and a 2-fold increase of sphingomyelin in PNS myelin.

Both proteins and lipids are asymmetrically distributed. Galactolipids are presumably located at the extracellular membrane surface (intraperiod line). It appears that there is twice as much cholesterol in the extracellular half of the bilayer as in the cytoplasmic half. In contrast, ethanolamine plasmalogen is localized to the cytoplasmic half of the bilayer. Proteins are more or less embedded in the bilayer and others attached to one surface or the other by weaker linkages (for review, see Kirschner et al 1984).

CNS myelin proteins

Most myelin proteins are unique to that structure and some of them are only present in CNS or PNS myelin (Table 2). CNS myelin has a relatively simple protein

Table 2   Myelin-specific proteins

	CNS		PNS	
	Oligodendrocyte	Myelin	Schwann cell	Myelin
MAG	(+)	+	(+)	+
CNP	+	(+)	(+)	0
MOG	+	+	0	0
P0	0	0	(+)	+
PLP/DM-20	(+)	+	(+)	0
MBP	(+)	+	(+)	+
P2	(spinal cord)		(+)	+
PMP-22	0	0	(+)	+

Low concentration (+); main localization +

composition: proteolipid protein (PLP) and myelin basic protein (MBP) make up 60–80% of the total. PLP is very hydrophobic, its sequence is highly conserved during evolution, and it spans several times the lipid bilayer. It has domains in both the intraperiod and major dense lines. PLP is probably involved in stabilizing the intraperiod line. MBP is located on the cytoplasmic face of the myelin membranes (major dense line) and plays a key role by keeping in close apposition the cytoplasmic faces of myelin lamellae. The enzyme 2′,3′-cyclic nucleotide 3′-phosphodiesterase (CNP) is enriched in both types of myelin-forming cells and their processes. The substrate for the enzyme does not exist in the nervous tissue and it is considered to be a structural protein involved in maintaining contacts between the cytoskeleton and the plasma membrane. Among the quantitatively minor myelin proteins, the myelin-associated glycoprotein (MAG) and the myelin/oligodendrocyte glycoprotein (MOG) are partially exposed at the external surface of the membrane: at the axolemma side for MAG and at the most external layer of the myelin sheath and oligodendroglial plasma membrane for MOG.

PNS myelin proteins

The protein composition of PNS myelin differs greatly from that of the CNS. PLP and MOG are absent and MBP concentration represents only approximately 10% of total protein. At least 50% of PNS myelin protein is represented by a glycoprotein called P0. It is a transmembrane protein with an extracellular domain containing the carbohydrate moeity and an intracellular domain (C-terminal). Indirect evidence suggests that P0 could play the role of PLP in the CNS, stabilizing the multilamellar structure of myelin. P0 is a member of the immunoglobulin superfamily; it contains a variable domain that, like other cell-recognition molecules (e.g. N-CAM), can make contact with another P0 molecule located on the neighbouring myelin lamella. This type of interaction is called homophilic. P2 is another basic protein, unrelated to MBP, that has also been observed in low concentration in the spinal cord, a CNS structure. It is not clear whether, in the spinal cord, P2 is produced by Schwann cells (producing PNS myelin) or by oligodendrocytes. MAG is present in PNS myelin at locations where myelin lamellae are separated by cytoplasm (paranodal loops, Schmidt–Lantermann clefts, etc.). Recently, a new integral PNS myelin protein, PMP-22, has been described (Suter et al 1992). It could be involved in the regulation of myelin growth.

All the known myelin proteins have now been cloned (for review see Lemke 1988) and their amino acid sequences determined. The use of mutant and transgenic animals has made it possible to understand the function of most of them. Furthermore, mutations in animals have been invaluable in interpreting the pathophysiological mechanisms of several genetic diseases in animals and humans.

MYELIN DISEASES

Definitions

Demyelination is the destruction of the normal myelin sheath. Dysmyelination indicates the formation of a sheath of abnormal composition. Its structure will be perturbed, unstable and usually rapidly degraded. Therefore, dysmyelination implies demyelination also. *Primary demyelination*, or better *primary myelin disease*, indicates that the primary site of injury is the myelin-forming cell and the myelin sheath, whereas neurons and axons are spared, at least in the early phase of the disease. A good example of primary demyelination is multiple sclerosis. In this disease typical plaques show a loss of oligodendrocytes while preserved axons are deprived of their myelin sheath. In contrast, *secondary demyelination* refers to myelin loss associated with Wallerian degeneration following axonal section or neuronal death. In this case it is implied that the neuron or its axon are the primary target of the lesion. In lesions devoid of inflammatory cells or blood cells (haemorrhage) myelin and its degradation products can persist for prolonged periods (Miklossy and Van der Loos 1991) and it is not clear why remyelination can occur in the PNS whereas it is always incomplete and inefficient in the CNS.

Classification

Classification of myelin diseases is still based on clinical signs and anatomopathological criteria. The improvement of our understanding of the pathophysiology of myelin diseases will make such classifications obsolete. Nevertheless, one can grossly divide myelin diseases according to their aetiology. Mainly diseases in which myelin is considered the primary target ('primary demyelination') will be considered (Table 3). This classification is a simplification of that of Raine (1984). In the future, inherited myelin diseases will be classified according to the molecular defect (genotype) rather than to the manifestation (phenotype).

In the following section four examples of primary myelin diseases and one secondary myelin disease are presented.

Pelizaeus–Merzbacher disease and the jimpy mouse

Jimpy is an X-linked mutation in mice that has been long considered as a murine form of inherited sudanophilic leukodystrophy. The assignment of the PLP gene to the X-chromosome suggested that a mutation in the PLP gene could be responsible for sex-linked myelin diseases. The *jimpy* mutation causes erroneous splicing of PLP mRNA, resulting in a truncated and abnormal C-terminal sequence. Several different point mutations have been identified in Pelizaeus–Merzbacher patients (Hudson et al 1990), giving full support to the old idea that the jimpy mouse was a murine model for this disease. Only oligodendrocytes and CNS myelin are affected, whereas PNS myelin appears normal. This was expected since PLP is not expressed in PNS myelin.

Table 3 Classification of primary myelin diseases

1. Allergic and infectious diseases
2. Hereditary metabolic diseases
3. Toxic diseases
4. Nutritional diseases
5. Traumatic diseases

For a more exhaustive list see Raine (1984)

Krabbe disease (globoid-cell leukodystrophy) and the twitcher mouse

This severe neurological disease is usually fatal during the second year of life. In the brain and the PNS of Krabbe patients there is an important loss of myelin and axons. Large rounded cells containing several nuclei and crystalloid cytoplasmic inclusions called globoid cells are typical of this disease. The defect is a deficiency of galactosylceramide β-galactosidase (Figure 2). The availability of a genetically authentic model of Krabbe disease, the *twitcher* mutant mouse provided the opportunity to study the pathophysiology of this disease. Suzuki and Suzuki (1990) demonstrated that in twitcher mice excess galactosylceramide (cerebroside) alone does not cause demyelination but the accumulation of psychosine (galactosylsphingosine) becomes highly cytotoxic and is responsible for the destruction of the myelin-forming cells and subsequent demyelination. These authors also showed that globoid cells, observed in Krabbe's patients and twitcher mice are macrophages altered by liberated excess of free galactosylceramide that cannot be degraded because of the enzymatic defect.

Charcot–Marie–Tooth disease and the trembler mouse

Trembler is an autosomal dominant mutation in mice with full penetrance affecting only the peripheral nerves. In trembler mice, Schwann cells have a high proliferation rate and are unable to maintain the myelin sheaths. Myelin sheaths are poorly compacted, with onion bulb formation of thinly myelinated axons with layers of redundant basal laminae. The *trembler* phenotype is very similar to the most common form of autosomal dominantly inherited hypertrophic neuropathy, Charcot–Marie–Tooth disease 1A. Charcot–Marie–Tooth (CMT) is one of the most common hereditary motor and sensory neuropathies, affecting roughly 1 in 2500 people. This

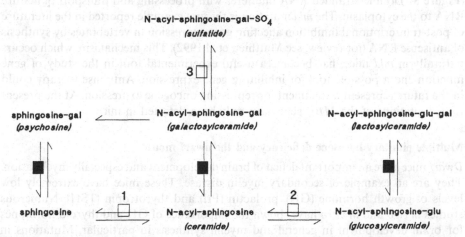

Figure 2 Metabolic pathways of galactosylceramide and related sphingolipids. The enzyme galactosylceramide β-galactosidase hydrolyses not only galactosylceramide but also psychosine and lactosylceramide. The metabolic block responsible for Krabbe disease is represented as a solid box. Open boxes represent defects of other enzymes responsible for: (1) Farber disease, (2) Gaucher disease, (3) metachromatic leukodystrophy

disease is caused by mutations of the p12–p11.2 region of human chromosome 17, which has evolutionarily conserved counterparts on mouse chromosome 11, near the *trembler* locus. Recently, Suter and colleagues (1992) have found in trembler mice a point mutation that substitutes an aspartic acid residue for a glycine in a putative membrane-associated domain of the PMP-22 protein. The gene of this peripheral myelin protein is assigned to chromosome 11. The evidence was sufficiently persuasive that Suter and colleagues (1992) predicted that *PMP-22* gene was a candidate gene for the CMT disorder. Several groups have now demonstrated that the *PMP-22* gene maps to the middle of the duplicated region in Charcot–Marie–Tooth 1A (Matsunami et al 1992; Patel et al 1992; Timmerman et al 1992). These findings demonstrate that trembler mice are a legitimate model for MCT not only on phenotypic grounds but also at the gene level. This is another demonstration of the value of mutant animals in investigating inherited diseases in humans.

Myelin-deficient, a mutation affecting myelin basic protein expression

Myelin-deficient (*mld*) mice are characterized by trembling of hind limbs and reduced life span. The few myelinated axons in *mld* brain are surrounded by thin uncompacted lamellae presenting an absence of major dense line in electron micrographs. This suggested that the primary defect was an impaired function or absence of myelin basic protein (MBP). Indeed *mld* mice have reduced MBP mRNA.

In *mld* mice the gene coding for MBP is tandem-duplicated. The upstream gene contains a large inversion of the 3′-region and therefore cannot give rise to mature mRNA and functional protein. The downstream gene is normal. Both genes are transcribed independently. The upstream gene containing the large inversion produces antisense RNA, which may form double-stranded RNA with sense primary transcript (Figure 3). Double-stranded RNA interferes with processing and transport of mature RNA to the cytoplasm. The *mld* mutation is the first example reported in the literature of post-transcriptional inhibition affecting gene expression in vertebrates by synthesis of antisense RNA (for review, see Matthieu et al 1992). This mechanism, which occurs naturally in *mld* mice, has become a useful experimental tool in the study of gene function and a possible tool for inhibiting gene expression. Antisense therapy could in the future represent a treatment for repressing oncogene expression. At the present time, mutations of the *MBP* gene have only been detected in mice.

Multiple pituitary hormone deficiency and the dwarf mouse

Dwarf mice have an important deficit of brain development and especially myelination. They are an example of secondary myelin disease. These mice have extremely low levels of growth hormone (GH), prolactin (Prl) and thyrotropin (TSH). Numerous studies *in vivo* and *in vitro* have shown the importance of GH and thyroid hormones for brain development in general and myelin synthesis in particular. Mutations in the gene encoding Pit-1, a member of the POU family of transcription factors, have been identified in Snell's dwarf mice (Li et al 1990). Pit-1 contains two protein domains, which are both necessary for high-affinity DNA binding on the genes encoding GH, Prl and TSH. *Dwarf* is very similar to multiple pituitary hormone

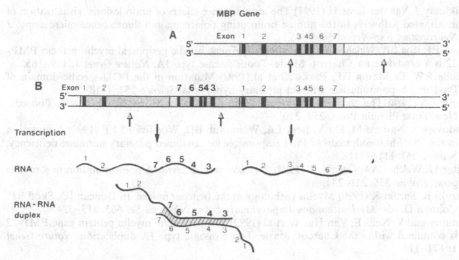

Figure 3 Organization of the myelin basic protein gene (*Mbp*) in wild-type and *mld* mutants. The formation of sense–antisense duplexes from two distinct transcripts is represented. The inverted DNA segment is delimited by open arrows. (A) Wild-type *Mbp* gene; (B) *Mbp*^mld allele

deficiency in humans. Furthermore, point mutations in a POU-specific portion of the human gene that encodes the pituitary-specific transcription factor Pit-1 have been identified in families affected by this hypopituitarism (Radovick et al 1992; Pfäffle et al 1992).

These few examples of mutations affecting myelin formation and maintenance in mice and humans emphasize the importance of animal models in studying the pathophysiology and molecular defects of inborn errors of metabolism in humans.

ACKNOWLEDGEMENT

I thank Jane Charlton for her assistance with this manuscript.

REFERENCES

Hudson LD (1990) Molecular genetics of X-linked mutants In Duncan ID, Skoff RP, Colman D, eds. *Myelination and Dysmyelination; Ann NY Acad Sci* **605**: 155–165.

Kirschner DA, Ganser AL, Caspar DLD (1984) Diffraction studies of molecular organization and membrane interactions in myelin. In Morell P, ed. *Myelin*, 2nd edn. New York: Plenum Press, 51–95.

Lemke G (1988) Unwrapping the genes of myelin. *Neuron* **1**: 535–543.

Li S, Crenshaw EB, Rawson EJ, Simmons DM, Swanson LW, Rosenfeld MG (1990) Dwarf locus mutants lacking three pituitary cell types result from mutations in the POU-domain gene *pit-1*. *Nature* **347**: 528–533.

Matsunami N, Smith B, Ballard L et al (1992) Peripheral myelin protein-22 gene maps in the duplication in chromosome 17p11.2 associated with Charcot–Marie–Tooth 1A. *Nature Genet* **1**: 176–179.

Matthieu J-M, Tosic M, Roach A (1992) Myelin deficient mutant mice. An *in vivo* model for inhibition of gene expression by natural antisense RNA. In Baserga DT, Denhardt DT, eds. *Antisense Strategies; Ann NY Acad Sci* **660**: 188–193.

Miklossy J, Van der Loos H (1991) The long distance effects of brain lesions: visualization of myelinated pathways in the human brain using polarizing and fluorescence microscopy. *J Neuropathol Exp Neurol* **50**: 1–15.

Patel PI, Roa BB, Welcher AA et al (1992) The gene for the peripheral myelin protein PMP-22 is a candidate for Charcot–Marie–Tooth disease type 1A. *Nature Genet* **1**: 159–165.

Pfäffle RW, Di Mattia GE, Parks JS et al (1992) Mutation of the POU-specific domain of Pit-1 and hypopituitarism without pituitary hypoplasia. *Science* **257**: 1118–1121.

Raine CS (1984) The neuropathology of myelin diseases. In Morell P, ed. *Myelin*, 2nd edn. New York: Plenum Press, 259–310.

Radovick S, Nations M, Du Y, Berg LA, Weintraub BD, Wondisford FE (1992) A mutation in the POU-homeodomain of Pit-1 responsible for combined pituitary hormone deficiency. *Science* **257**: 1115–1118.

Suter U, Welcher AA, Özcelik T (1992) *Trembler* mouse carries a point mutation in a myelin gene. *Nature* **356**: 241–244.

Suzuki K, Suzuki K (1990) Myelin pathology in the twitcher mouse. In Duncan ID, Skoff RP, Colman D, eds. *Myelination and Dysmyelination*; *Ann NY Acad Sci* **605**: 313–324.

Timmerman V, Nelis E, Van Hul W et al (1992) The peripheral myelin protein gene PMP-22 is contained within the Charcot–Marie–Tooth disease type 1A duplication. *Nature Genet* **1**: 171–175.

J. Inher. Metab. Dis. 16 (1993) 733–743

The Inherited Leukodystrophies: A Clinical Overview

J. AICARDI

Hôpital des Enfants Malades, Rue de Sèvres 149, F-75743 Paris Cedex 15, France

Summary: The leukodystrophies are degenerative diseases that involve primarily the white matter of the brain. The most common leukodystrophies result from known disturbances in the synthesis or catabolism of myelin such as a block in the catabolism of sulphatides and of galactocerebrosides, respectively, in metachromatic leukodystrophy and in Krabbe disease, or from synthesis of an abnormal proteolipid protein in Pelizaeus–Merzbacher disease. The cause of white matter involvement in other leukodystrophies remains unknown even though metabolic anomalies, such as accumulation of acetylaspartic acid in Canavan disease, have been demonstrated.

Common clinical features of the leukodystrophies include neurological deterioration following a period of normal development, predominant involvement of motor function at least initially, and absence of convulsions or myoclonus. Imaging – especially magnetic resonance – shows changes in density or signal from central white matter.

Most leukodystrophies feature suggestive symptoms and signs such as effects on peripheral nerves' myelin in Krabbe disease and metachromatic leukodystrophy, or X-linked inheritance and slow deterioration in Pelizaeus–Merzbacher disease.

Therapy of the leukodystrophies is purely symptomatic in most cases. Trials of bone marrow transplantation are being pursued for metachromatic leukodystrophy and adrenoleukodystrophy.

The leukodystrophies are degenerative diseases that involve primarily and preferentially the white matter of the brain. Such a definition is vague and includes a large and heterogeneous group of disorders of different mechanisms, courses, and clinical and pathological features that are related only by the localization of the pathological process and by their progressive, degenerative nature and genetic character. The limits of the group are imprecise and vary with the criteria used for inclusion. For example, severe and progressive involvement of the white matter occurs in disorders of the metabolism of amino acids such as phenylketonuria (Bauman and Kemper 1982; Pearsen et al 1990), or of organic acids such as glutaric aciduria type I (Aicardi 1992), or in peroxisomal disorders in which the primary process is not limited to the white matter. Likewise, the degree of myelin loss and gliosis in Tay–Sachs disease often seems to be more pronounced than could be expected from simple neuronal loss, suggesting a direct effect on the white matter (Adams and Lyon 1982). Conversely, in undoubted leukodystrophies such as sulphatidosis, neurons are often involved and specific inclusions are found in perikarya.

Clinical and imaging criteria allow for the recognition of a group of disorders that are marked by an abnormal signal from the white matter on CT scanning or MR imaging, usually in association with symptoms and signs of diffuse CNS dysfunction, which can be termed leukoencephalopathies but clearly include many conditions, some non-progressive or non-genetic that do not fit the definition of the leukodystrophies. Pathological criteria are more precise and permit separation of cases with isolated or predominant extensive demyelination or absence of myelin from those which diffusely affect the brain although it may be difficult in some cases to determine whether the degree of myelin disorder is entirely due to disease of neuronal perikarya or is a primary phenomenon.

Traditional pathological distinctions between cases with metachromatic as opposed to orthochromatic myelin breakdown products, or with patchy as opposed to diffuse myelin loss, are still used in classification of the leukodystrophies but are not relevant to the definition of these conditions.

Biochemical and enzymatic studies and, more recently, DNA analysis have allowed precise characterization of the myelin disorder in some cases but are not universally applicable. It is now clear that different forms of leukodystrophy may result not only from lysosomal enzyme abnormalities affecting the catabolism of myelin but also from deficient synthesis of myelin, or from disturbances of metabolic pathways whose relationship to myelin synthesis or maintenance is imperfectly understood, as is the case in adrenoleukodystrophy or acylaspartase deficiency. Thus, biochemical disturbances do not permit a meaningful classification. Moreover, there appears to be no compelling reason for not including disorders such as phenylketonuria, as there may be extensive and progressive demyelination in this disease which appears not to be secondary to disease of the perikaryon (Meydig-Lamade et al 1992).

A list of the main leukodystrophies appears in Table 1, which also indicates the corresponding enzymatic or other biochemical abnormality and the mode of genetic transmission. The term primary leukodystrophy refers to cases in which the leukodystrophy is an isolated or a predominant feature. Secondary leukodystrophies are those conditions in which the clinical features of the white matter disease are overshadowed by other neurological and/or systemic manifestations. Most diseases are transmitted as autosomal recessive traits. However, Pelizaeus–Merzbacher disease and the late form of adrenoleukodystrophy are X-linked diseases.

This review deals primarily with those leukodystrophies in which a disorder in the synthesis or metabolism of myelin has been documented, as these are the best-defined members of the group.

Other genetic leukodystrophies associated with known metabolic disturbances of the metabolism of substances that are probably related to the synthesis or maintenance of myelin, such as very long-chain fatty acids and acetylaspartic acid, and those that affect metabolic pathways unrelated to myelin, for example DNA repair in Cockayne disease, will be dealt with very briefly. The leukodystrophies that are only secondary phenomena, such as that encountered in phenylketonuria patients, will not be discussed.

Table 1 Main leukodystrophies

Disorder	Mode of inheritance	Metabolic abnormality
I. **Primary leukodystrophies with known metabolic defect**		
Metachromatic leukodystrophy	AR	Cerebroside sulphatase
Krabbe disease	AR	β-Galactocerebrosidase
Multiple sulphatase deficiency	AR	Arylsulphatase A,B,C; MPS sulphatase
X-linked adrenoleukodystrophy	XLR	VLCFA-CoA synthetase
Neonatal adrenoleukodystrophy and Zellweger syndrome	AR	Absence of peroxisomes
Canavan–Van Bogaert disease	AR	Acylaspartase
Pelizaeus–Merzbacher disease	XLR	Deficient PLP synthesis
II. **Primary leukodystrophies without known metabolic defect**		
Sudanophilic (orthochromatic) leukodystrophies	AR	
Alexander disease (leukodystrophy with Rosenthal fibres)	?AR	
Leukodystrophy with calcification of basal ganglia and CSF lymphocytosis	AR	
III. **Leukodystrophies secondary to other metabolic disorders**		
Mucopolysaccharidoses, mainly Hurler disease	AR	L-Iduronidase
Phenylketonuria	AR	Phenylalanine oxidase
Glutaric aciduria	AR	Glutaryl-CoA dehydrogenase
Other amino acid or organic acid disorders	AR	
Mitochondrial encephalopathies and encephalomyopathies	AR or mito	COX, complexes I, II, III
Cockayne disease	AR	

AR = autosomal recessive; mito = mitochondrial (maternal) inheritance; XLR = X-linked recessive; COX = cytochrome *c* oxidase; MPS = mucopolysaccharides; PLP = proteolipid protein; VLCFA = very long-chain fatty acids

THE CLINICAL CONCEPT OF LEUKODYSTROPHY

The conditions grouped under the heading of leukodystrophy are so heterogeneous that it is doubtful whether such a heading is practically useful. However, the leukodystrophies share some common features, even though exceptions are frequent. The clinical picture is dominated by motor manifestations that include spasticity, weakness, pyramidal tract signs and cerebellar signs. Such signs are usually bilaterally symmetrical and slowly progressive. They are due to diffuse dysfunction of axonal conduction as a result of demyelination, which affects mainly the long myelinated tracts such as pyramidal and spinocerebellar fascicles. Spastic diplegia, often with some degree of associated ataxia, is especially suggestive. Epileptic seizures, myoclonic attacks or erratic myoclonus and paroxysmal EEG abnormalities are absent or appear only late in the course. This is related to the absence of direct effects on perikarya so that the electrical properties of their membranes are not modified. However, the occurrence of seizures in some leukodystrophies such as Krabbe disease or adrenoleukodystrophy may be due to the extension of lesions to the grey matter, which is known to occur in a significant proportion of cases (Adams and Lyon 1982). Optic atrophy is often present but retinal degeneration with involvement of sensory epithelium is lacking. Cognitive and behavioural deterioration may appear at any time but are often late and tend to be overshadowed by motor abnormalities. These characteristics contrast with those of the poliodystrophies that are marked by early mental deterioration, epilepsy or myoclonus with only secondary motor involvement.

This traditional picture is of only limited value as a result of the frequent deviations from this schematic picture. For example, seizures are not infrequently an early manifestation of adrenoleukodystrophy and EEG abnormalities and myoclonic attacks are common in infantile Krabbe leukodystrophy. Likewise, early cognitive deterioration and/or psychiatric disturbances are among the most common manifestations of juvenile metachromatic leukodystrophy. In practice, the diagnosis in most cases of leukodystrophy is usually made because individual types exhibit specific clinical profiles that may be quite different from that traditionally ascribed to the group as a whole. It is therefore necessary to outline briefly such profiles at least for the most common forms of leukodystrophy. The overall clinical features of the leukodystrophies in general are only rarely useful and permit at best a tentative assignment of unclassifiable degenerative CNS diseases to a probable lesional localization, which may be of some help in planning further investigations.

THE DIAGNOSIS OF THE INHERITED LEUKODYSTROPHIES

Despite multiple possible clinical presentations, the leukodystrophies share some important features that permit categorization among the genetic metabolic diseases and differentiation from other, non-progressive CNS diseases in the same age range.

The most important clue to the diagnosis of degenerative and metabolic diseases is the progressive nature of the disorder. In practice, almost all progressive diseases are degenerative and genetically transmitted and the most important clinical decision is to establish or exclude progressivity (Adams and Lyon 1982). The notable exceptions to this general rule are inflammatory diseases that can closely mimic metabolic disorders. These include some forms of chronic encephalitis and AIDS encephalopathy (Aicardi 1989), the latter proving to be a frequent cause of progressive encephalopathies in some settings.

The progressive character is not always easily affirmed. Two essential arguments are the notion of a symptom-free interval during which initial development has been normal and the loss of already-acquired skills. The main conditions to be differentiated are: (1) static or non-progressive psychomotor retardation; (2) seizure states and the effects of anti-epileptic medication; and (3) sociocultural retardation. The diagnosis of non-progressive diseases may be difficult in very young patients for whom regression may be impossible to assess, in patients receiving heavy sedative drug treatment, in patients with severe systemic disorders that make neurodevelopmental assessment unreliable, and in those in whom deterioration sets in insidiously over several months or years or in whom initial development has never been normal. In addition, some forms of actual deterioration do not indicate a relentlessly progressive organic brain disease. Loss of acquired skills occurs with epilepsy, especially with West and Lennox–Gastaut syndromes, but deterioration stops after some time and resumption of progress often takes place. In some cases of autism, there is no apparent brain pathology despite clear deterioration (Rapin 1991). Finally, the emergence of new neurological signs need not be proof of degenerative disease. It is often observed in infants with cerebral palsy, e.g. congenital hemiplegia or dystonia, especially in the first years of life, despite stability of the pathological lesions (Adams and Lyon 1982; Aicardi 1989).

Once a progressive disease is recognized, the next step is to rule out non-metabolic causes. Some cases of brain tumours, hydrocephalus, vascular disorders responsible for repeated strokes, chronic ischaemia due to the presence of a 'blood steal' through an arteriovenous malformation, and inflammatory and infectious diseases can be responsible and are often more amenable to effective treatment. Exclusion of these conditions as well as that of metabolic diseases other than the leukodystrophies regularly requires the use of neuroimaging techniques, especially magnetic resonance studies. The choice of further tests will vary as a function of both the clinical and neuroimaging presentations.

Tests that give useful indications in favour of a diagnosis of leukodystrophy – in addition to specific biochemical and enzymatic tests mentioned below – include an EEG that shows progressive slowing of background rhythms and does not generally contain paroxysmal episodes, and evoked potentials (visual, auditory or somatosensory) that are regularly abnormal in shape and of increased latency as a result of delayed conduction while travelling through the white matter despite relative normality of receiving cortical areas (Taylor and Fagan 1990). The electroretinogram is useful mainly to exclude involvement of the sensory retinal epithelium, which is only exceptionally involved in the leukodystrophies, in sharp contrast to what often obtains in some poliodystrophies.

Nerve conduction velocities are low in three diseases – metachromatic leukodystrophy and its variants, Krabbe disease and Cockayne disease – because of marked demyelination in peripheral nerves. The presence of a neuropathy is of value for the diagnosis of the type of leukodystrophy.

Biopsy, for examination of peripheral tissue, skin, conjunctiva and rectal mucosa is rarely necessary for most leukodystrophies and brain biopsy is hardly justified today, unless it appears to be the only way of demonstrating an inflammatory disease.

Imaging is a major tool for the diagnosis of the leukodystrophies. The finding of symmetrical areas of hypodensity on CT scanning often permits a rapid diagnosis of leukodystrophy and sometimes suggests a particular form as some patterns are fairly specific. However, typical features are not present in all cases. Conversely, similar areas of hypodensity may be associated with many other progressive or non-progressive disorders such as mitochondrial myopathies or post-hypoxic leukoencephalopathy. MRI is much more sensitive than CT and does not fail to show demyelination, when present, as a low signal from the white matter on T_1-weighted images even when CT does not show any parenchymal abnormality. However, absent or insufficient myelin is also demonstrated in other conditions. Imaging findings are discussed in detail elsewhere (Kendall 1993) and some peculiarities of certain forms are briefly indicated below.

INDIVIDUAL FEATURES OF SOME LEUKODYSTROPHIES

A few individual clinical features and some important laboratory features are described only for the main leukodystrophies that are due to a known enzymatic abnormality in the degradation of myelin (metachromatic leukodystrophy and Krabbe disease) or to defective synthesis (Pelizaeus–Merzbacher disease). For the remaining diseases, only a few essential characteristics that can help the diagnosis are given.

Leukodystrophies due to abnormal myelin synthesis or degradation

Metachromatic leukodystrophy: MLD, or sulphatidosis, is an autosomal recessive disease caused by deficiency of cerebroside sulphatase, a protein coded for by a gene on chromosome 22. An exceptional variant is due to deficiency in cerebroside sulphatase activator, coded for by a gene on chromosome 10. In practice, artificial substrates are used that test the arylsulphatase A activity. Myelin breakdown products are characteristically metachromatic, i.e. the tissue–dye complex has an absorption spectrum sufficiently different from the original dye to produce an obvious change in colour. The spectral shift results from polymerization of the dye under the influence of negative charges such as sulphate residues closely spaced in tissue.

There are three phenotypes: the infantile form (1 : 40 000 population), the juvenile form (1 : 160 000), and the adult form. These phenotypes result from different DNA mutations. Infantile forms are due to homozygosity for an allele that does not direct the synthesis of polypeptides with arylsulphatase activity, whereas allele A, associated with the adult phenotype, codes for an unstable polypeptide with a reduced half-life. The juvenile phenotype is associated with compound heterozygosity for alleles A and I (Gieselman et al 1991; Polten et al 1991; Kappler et al 1992). In addition, a pseudodeficiency allele is responsible for very low arylsulphatase A activity in a significant proportion of the population. Compound heterozygotes for the pseudodeficiency and an MLD allele may have variable degrees of involvement (Gieselman 1991; Kappler et al 1991). A level of arylsulphatase of 10% or more is unassociated with phenotypic manifestations. Prenatal diagnosis is feasible on cultured amniocytes or on chorionic villi. Because of the frequency of the pseudodeficiency gene in the population (7.3–15%), prenatal diagnosis requires knowledge of the enzymatic structure of both parents since a low activity may be found in asymptomatic individuals carrying the pseudo-deficiency allele.

The clinical features of MLD in the infantile form are usually typical for a white matter disease, with spastic diplegia often associated with some degree of ataxia. Involvement of peripheral nerves is present in almost all cases so that there is usually reduction or abolition of deep tendon reflexes and markedly decreased motor and sensory nerve conduction velocities (MacFaul et al 1982). In rare cases, peripheral nerve involvement is the sole clinical manifestation for periods of weeks, thus masquerading as polyradiculoneuropathy. The picture of the juvenile form is usually dominated by mental regression and/or behavioural disturbances.

CT scans or MRI in all forms show white matter changes that may initially be slight. Diagnosis is by demonstration of low arylsulphatase activity. The presence of sulphatides in urine is a useful confirmation of the enzymatic block, but peripheral nerve or skin biopsies that show typical storage are no longer indicated.

Austin disease: Austin disease, or multiple sulphatase deficiency, shares with MLD a deficiency of arylsulphatase A, but arylsulphatase B and C, steroid sulphatase and mucopolysaccharide sulphatases are also deficient (Burch et al 1986; Aicardi 1992). The basic defect may be that of a regulatory gene that controls the metabolism of the various sulphatases. The hallmarks of this exceptional condition are coarse features, hepatomegaly, skeletal changes and icthyosis in addition to the features of MLD.

Krabbe leukodystrophy: This is remarkable by its early onset in the most common early infantile form. Involvement of peripheral nerves is also present but, in contrast to MLD, seizures and EEG paroxysms are frequent. Restlessness, irritability and progressive stiffness are major and early features (Hagberg 1984) but severe hypotonia is a known variant. Pathologically, there is marked atrophy of the brain. Large multinucleated cells (globoid cells) are seen in the demyelinated areas and store a protein-bound cerebroside.

The diagnosis rests on demonstration of a profound deficit in galactocerebroside-β-galactosidase activity. Interestingly, CT scan often shows not hypodensity of the white matter but signs of brain atrophy and zones of increased attenuation in the deep grey matter and in the periventricular and capsular white matter (Sasaki et al 1991). MRI shows extensive demyelination.

Juvenile Krabbe disease has more variable features. Visual failure is often prominent, while peripheral neuropathy and hyperproteinorrhachia are often absent (Fiumara et al 1990; Kolodny et al 1991; Lyon et al 1991; Phelps et al 1991). Although the course is rapid in the infantile form, a slow progression may be observed in the late types and a partial remission has been observed in one case (Goebel et al 1990).

Pelizaeus–Merzbacher disease: This is transmitted as an X-linked trait. It is the best-defined of the orthochromatic leukodystrophies. Patchy demyelination with the presence of preserved islands of myelin, the so-called 'tigroid pattern' is not specific for the disease and is absent in the most severe forms. Little sudanophilic material is present in the white matter. Oligodendrocytes contain cytoplasmic inclusions and numerous myelin balls at their periphery, a picture that suggests impairment in myelin formation rather than a destructive process. A defect in the synthesis of proteolipid protein (PLP), coded for by a gene assigned to the long arm of the X chromosome at Xq21-2 Xq22, is suggested by the similarities with the situation in the 'jimpy' mutation of mice, which have a mutant allele at the PLP locus. Indeed, interstitial deletions (Koeppen et al 1987) and point mutations in several exons (Gencic et al 1989; Pratt 19991; Boespflug-Tanguy et al 1992) have been demonstrated in some cases of Pelizaeus–Merzbacher disease. In most cases, however, no DNA anomaly is detectable, but in several families a tight linkage was observed with the cDNA PLP probe and two polymorphic probes located at Xq21.3 (Boespflug-Tanguy et al 1992). In informative pedigrees a prenatal diagnosis may thus be possible and the use of other polymorphic probes around and inside the PLP gene will shortly improve carrier detection and prenatal diagnosis.

From a clinical point of view, PMD is remarkable by its early onset with nystagmus, at or shortly after birth, and by a slow course that is compatible in many patients with survival into adult life with variable degrees of mental progress. The severe connatal form of Seitelberger disease, which leads to early death, is not genetically distinct from the later form as both types often coexist in the same lineage (Boespflug-Tanguy et al 1992). Because of slow course and apparent lack of progression, the misdiagnosis of static diseases such as spinocerebellar degenerations is often made. The most useful test is MRI evidencing massive demyelination, although CT scan is usually not diagnostic. There is no peripheral nerve involvement.

Leukodystrophies not due to a known defect in myelin metabolism

Adrenoleukodystrophy: Adrenoleukodystrophy is not a single disease but includes at least two different disorders with different modes of genetic transmission. The most common is the X-linked form (Moser 1989; Moser et al 1991). The clinical features of this disease often differ widely from those commonly attributed to the leukodystrophies. In particular, convulsions are not rare and may be severe and localized. Cortical visual impairment and sometimes central auditory deafness reflect the usual posterior predominance of demyelination that is a major diagnostic feature on CT scan and MRI. High intracranial pressure is not uncommon. This may be because of the intensity of inflammatory phenomena that are also responsible for the marked enhancement seen on imaging at the periphery of demyelinated areas. Symptoms and signs of adrenal dysfunction are usually minimal but the finding of pigmentation, sometimes localized only to scars, may be highly suggestive. The neonatal form of adrenoleukodystrophy is very different from the X-linked type as well as from other leukodystrophies and will not be considered. The leukodystrophy of Zellweger syndrome is closely related. However, in both types, the diagnosis can be confirmed easily by demonstrating the increased level of very long-chain saturated fatty acids, especially an elevated ratio of $C_{26}:C_{22}$ acids (Moser 1989).

Other leukodystrophies: Two rare leukodystrophies are clinically remarkable by the presence of macrocephaly: *Canavan-Van Bogaert disease* (or spongy degeneration of the CNS) and *Alexander disease*, or leukodystrophy with Rosenthal fibres. Their clinical presentation is otherwise completely different. Spongy degeneration runs a rapid course and is rapidly lethal, whereas the course of Alexander disease is usually slow (Borrett and Becker 1985; Gascon et al 1990). The genetics of Alexander disease are unclear. Most pathologically verified cases have been isolated ones. However, cases of leukodystrophy with large heads and familial occurrence are on record. In some of these (Harbord et al 1990) neither Rosenthal fibres nor spongiosis were present.

Another clinically distinctive leukodystrophy is *Cockayne disease*, in which progeric features, failure to thrive and slowly progressive neurodevelopmental deficit become apparent from the second year of life. Ventricular dilatation and intracranial calcification are prominent features and peripheral nerves are affected as shown by decreased conduction velocities.

Leukodystrophy with chronic CSF lymphocytosis and calcifications of the basal ganglia or Aicardi–Goutières syndrome (Aicardi and Goutières 1984) is a poorly defined leukodystrophy of neonatal or very early onset and a severe course with little or no developmental progress. Intense stiffness with opisthotonic attacks is described. CSF pleiocytosis persists for several months.

MANAGEMENT AND PREVENTION

No curative treatment is currently available for the leukodystrophies. For adrenoleukodystrophy, a dietary therapy combining restriction of long-chain fatty acids and administration of oleic acid has been found to be effective in normalizing accumulation of very long-chain fatty acids (Rizzo et al 1987; Moser 1989; Uziel et al 1991). There is some evidence that progression of the disease can be stopped, but any severe neurological abnormality already present is unlikely to improve, so that therapy may not be justified in advanced cases. Dietary treatment of presymptomatic children is indicated and seems to be effective (Uziel et al 1991). Bone marrow transplantation has been shown to reverse early neurological and neuroradiological dystrophy (Aubourg et al 1990), but this therapy is still in the experimental stage and the balance of risks should be carefully assessed. Transplantation in clinically presymptomatic children may be indicated if early MRI abnormalities appear, usually as small symmetrical areas of increased T_2 signal in the internal capsule or centrum semiovale (Aubourg et al 1989). Treatment of any manifestation of adrenal insufficiency is in order. ACTH has been reported to bring about partial improvement in rare cases, perhaps by modifying the intense immune response that is present in this disease.

Bone marrow transplantation has also been attempted in metachromatic leukodystrophy and some improvement has been reported (Krivit et al 1990), but more experience is clearly needed as sulphatides begin to accumulate long before birth and penetration of the enzyme in the CNS is questionable.

Prevention is based on precise and early diagnosis of affected children, genetic counselling and prenatal diagnosis. Detection of heterozygotes is possible by determination of very long-chain fatty acid levels in adrenoleukodystrophies and, in some families of Pelizaeus–Merzbacher disease, by linkage analysis. A reliable prenatal diagnosis is available for all leukodystrophies with known enzymatic defect using either amniocytes or chorionic villus biopsy (Table 1). No prenatal diagnosis is possible in leukodystrophies without known metabolic defect except in some cases of Pelizaeus–Merzbacher disease.

CONCLUSION

The term leukodystrophy has several possible definitions, varying from those that include all genetic diseases of the white matter to those that only accept a primary disorder of myelin synthesis or catabolism. The clinical concept of leukodystrophy (as opposed to grey matter diseases) is of doubtful usefulness as individual entities within the group do not necessarily share common features and clinical diagnosis is basically by identification of clinical profiles specific for each entity. Neuroimaging and biochemical techniques have considerably increased our diagnostic abilities and understanding of the leukodystrophies, but therapeutic possibilities remain quite limited.

REFERENCES

Adams RD, Lyon G (1982) *Neurology of Hereditary Metabolic Diseases of Children.* New York: McGraw-Hill.

Aicardi J (1989) Progrediente Enzephalopathie-Syndrome unbekannter Ursache. In Hanefeld F, ed. *Aktuelle Neuropädiatrie.* Heidelberg: Springer Verlag, 3–15.

Aicardi J (1992) *Diseases of the Nervous System in Childhood.* London: MacKeith Press.

Aicardi J, Coutières F (1984) A progressive familial encephalopathy in infancy with calcifications of the basal ganglia and chronic cerebrospinal fluid lymphocytosis. *Ann Neurol* **16**: 60–65.

Aubourg PR, Sellier N, Chaussain JL, Kalifa G (1989) MRI detects cerebral involvement in neurologically asymptomatic patients with adrenoleukodystrophy. *Neurology* **39**: 1619–1621.

Aubourg PR, Blanche S, Jambaqué I et al (1990) Reversal of early neurologic and neuroradiologic manifestations of X-linked adrenoleukodystrophy by bone marrow transplantation. *N Engl J Med* **322**: 1860–1866.

Bauman ML, Kemper TL (1982) Morphologic and histoanatomic observations of the brain in untreated phenylketonuria. *Acta Neuropathol* **58**: 55–63.

Boespflug-Tanguy O, Cavagna A, Mimault C et al (1992) X-linked inherited demyelinating disease (Pelizaeus–Merzbacher disease); tight linkage with the proteolipoprotein gene (PLP) in 17 families. *J Neurol* **239** (Suppl.2): S7.

Borrett D, Becker LE (1985) Alexander disease; A disease of astrocytes. *Brain* **108**: 367–385.

Burch M, Fensom AH, Jackson M, Pitts-Tucker T, Congdon PJ (1986) Multiple sulphatase deficiency presenting at birth. *Clin Genet* **30**: 409–415.

Fiumara A, Pavone L, Siciliano L, Tine A, Parano E, Innico G (1990) Late-onset globoid cell leukodystrophy: report on seven more patients. *Childs Nerv Syst* **6**: 194–197.

Gascon CG, Ozand PT, Mahdi H, Jamil A, Haider A, Brismar J (1990) Infantile CNS spongy degeneration in 14 cases: clinical update. *Neurology* **40**: 1876–1882.

Gencic S, Abuelo O, Ambler M, Hudson LD (1989) Pelizaeus–Merzbacher disease: an X-linked neurologic disorder of myelin metabolism with a novel mutation in the gene encoding proteolipid protein. *Am J Hum Genet* **45**: 435–442.

Gieselman V (1991) An assay for the rapid detection of the arylsulfatase A pseudo-deficiency allele facilitates the diagnosis and the genetic counselling for metachromatic leukodystrophy. *Hum Genet* **86**: 251–255.

Gieselman V, Polten A, Kreysing J, Kappler J, Fluharty A, Von Figura K (1991) Molecular genetics of metachromatic leukodystrophy. *Dev Neurosci* **13**: 222–227.

Goebel HH, Harzer K, Ernst JP, Bohl J, Klein H (1990) Late-onset globoid cell leukodystrophy: unusual ultrastructural pathology and subtotal beta-galactocerebroside deficiency. *J Child Neurol* **5**: 299–307.

Hagberg B (1984) Krabbe disease: clinical presentation of neurological variants. *Neuropediatrics* **15** (Suppl): 11–15.

Harbord MG, Harden P, Harding B, Brett EM, Baraitser M (1990) Megalencephaly and dysmyelination, spasticity, ataxia, seizures and distinctive neurophysiological findings in two siblings. *Neuropediatrics* **21**: 164–168.

Kappler J, Watts RWE, Conzelmann E, Gibbs DA, Propping G, Gieselmann V (1991) Low arylsulfatase A activity and choreoathetotic syndrome in three siblings: differentiation of pseudodeficiency from metachromatic leukodystrophy. *Eur J Pediatr* 150: 287–290.

Kappler J, Von Figura K, Gieselman V (1992) Late-onset metachromatic leukodystrophy: molecular pathology in two siblings. *Ann Neurol* 31: 256–261.

Kendall BE (1993) Inborn errors and demyelination: MRI and the diagnosis of white matter disease. *J Inher Metab Dis* 16: 771–786.

Koeppen A, Ronca NA, Greenfield EH, Hans KB (1987) Defective biosynthesis of proteolipid protein in Pelizaeus–Merzbacher disease. *Ann Neurol* 21: 159–170.

Krivit W, Shapiro E, Kennedy W et al (1990) Treatment of late infantile metachromatic leukodystrophy by bone marrow transplantation. *N Engl J Med* 322: 28–32.

Kolodny EH, Raghavan S, Krivit W (1991) Late-onset Krabbe disease (globoid-cell leukodystrophy): clinical and biochemical features of 16 cases. *Dev Neurosci* 13: 232–239.

Lyon G, Hagberg B, Evrard P, Allaire C, Pavone L, Vanier M (1991) Symptomatology of late onset Krabbe leukodystrophy: the European experience. *Dev Neurosci* 13: 240–244.

MacFaul R, Cavanagh N, Lake BD, Stephens R, Whitfield AE (1982) Metachromatic leukodystrophy: review of 38 cases. *Arch Dis Child* 57: 168–175.

Meyding-Lamade UK, Pietz J, Fahrendorf G, Sartor KJ (1992) Early treated phenylketonuria: do MR images parallel clinical, biochemical and neurophysiological data? *J Neurol* 239 (Suppl 2): S5.

Moser HW (1989) Peroxisomal disease. *Adv Pediatr* 36: 1–38.

Moser HW, Moser AB, Naidu S, Bergin A (1991) Clinical aspects of adrenoleukodystrophy and adrenomyeloneuropathy. *Dev Neurosci* 13: 254–261.

Pearsen KD, Gean-Marton AD, Levy HL, Davis K (1990) Phenylketonuria: MR imaging of the brain with clinical correlation. *Radiology* 177: 437–440.

Phelps M, Aicardi J, Vanier M (1991) Late-onset Krabbe leukodystrophy. A report of four cases. *J Neurol Neurosurg Psychiatr* 54: 293–296.

Polten A, Fluharty AL, Kappler J, Von Figura K, Gieselman V (1991) Molecular basis of different forms of metachromatic leukodystrophy. *N Engl J Med* 324: 18–22.

Pratt VM, Troffater JA, Schinzel A, Dlonhy SR, Conneally PM, Hodes ME (1991) A new mutation in the proteolipid protein (PL) gene in a German family with Pelizaeus–Merzbacher disease. *Am J Med Genet* 38: 136–139.

Rapin I (1991) Autistic children: diagnostic and clinical features. *Pediatrics* 87: 751–760.

Rizzo WB, Philipps MW, Dammann AL et al (1987) Adrenoleukodystrophy: dietary oleic acid lowers hexacosanoate levels. *Ann Neurol* 21: 232–239.

Sasaki M, Sakuragawa N, Takashima S, Hanaoka S, Arima M (1991) MRI and CT findings in Krabbe disease. *Pediatr Neurol* 7: 283–288.

Taylor MJ, Fagan ER (1990) Somatosensory evoked potentials: a review of their application in paediatric neurology. In Galai V, ed. *Maturation of the CNS and Evoked Potentials.* Amsterdam: Elsevier, 119–124.

Uziel G, Bertini E, Bardelli P, Rimoldi M, Gambetti M (1991) Experience on therapy of adrenoleukodystrophy and adrenomyeloneuropathy. *Dev Neurosci* 13: 274–279.

J. Inher. Metab. Dis. 16 (1993) 744–752
© SSIEM and Kluwer Academic Publishers. Printed in the Netherlands

Canavan Disease: Biochemical and Molecular Studies

R. MATALON, R. KAUL and K. MICHALS
The Research Institute Miami Children's Hospital, 6125 S.W. 31st Street, Miami, FL33155, USA

Summary: Deficiency of the enzyme aspartoacylase and the accumulation of *N*-acetylaspartic acid lead to a severe leukodystrophy and spongy degeneration of the brain, Canavan disease (McKusick 271900). Since our discovery in 1988 of the defect in Canavan disease, 144 patients with Canavan disease have been diagnosed in our laboratory. Most of these children are of Ashkenazi Jewish extraction. The level of enzyme activity can be used for carrier testing. Prenatal diagnosis has been difficult using the enzyme assay owing to the low activity of aspartoacylase in cultured chorionic villus samples or amniocytes. The determination of *N*-acetylaspartic acid in the amniotic fluid is another parameter for diagnosis; however, the levels may not always be elevated.

Bovine and human aspartoacylase have been purified in our laboratory. Bovine and human cDNA and genomic clones have been isolated and six exons have been localized. This information is being used for the study of Canavan disease at the molecular level.

N-Acetylaspartic aciduria was first described by Kvittingen and coworkers (1986) and then by Hagenfeldt et al (1987). Their investigations suggested a new organic acid disorder. The deficiency of aspartoacylase (EC 3.5.1.15) and *N*-acetylaspartic aciduria in Canavan disease was established in three patients in 1988 (Matalon et al 1988). These studies were later expanded to include a larger number of affected patients with Canavan disease and enzyme activity of obligate carriers (Matalon et al 1989, 1990).

Canavan disease, also known as spongy degeneration of the brain, was first described in 1928 by Globus and Strauss and in 1931 by Canavan. These authors thought they were dealing with Schilder disease. A detailed description of spongy degeneration of the brain as we understand it today was reported in three Jewish infants by van Bogaert and Bertrand in 1949. Reports that followed suggested that the disease is inherited as an autosomal recessive disorder with higher prevalence among people of Ashkenazi Jewish origin (Adachi et al 1973; Banker et al 1964; van Bogaert et al 1967; Buchanan et al 1965; Gambetti et al 1969; Menkes 1985; Sacks et al 1965; Ungar et al 1983). Banker and Victor (1979) described 48 Canavan patients, of whom 28 were Jewish, from a location in the vicinity of Vilnius in Lithuania. In spite of the prevalence of this disease among Jewish individuals, Canavan disease has also been described among other ethnic groups.

The diagnosis of Canavan disease can be suggested by the clinical features, which include leukodystrophy, megalencephaly, mental retardation and optic atrophy. Death usually occurs in the first decade of life. Further studies to substantiate the diagnosis would include CT scan or MRI showing white matter attenuation. Rushton and colleagues (1981) suggested that the CT scan in Canavan patients is different from patients with other leukodystrophies and megalencephaly. However, the most confirmatory and reliable diagnostic measure prior to the discovery of the enzyme defect was a brain biopsy showing spongy degeneration. The histopathology of the brain in Canavan disease is striking, with spongy degeneration of the myelin fibre, astrocytic swelling and elongated mitochondria. In spite of these specific findings there have been reports suggesting that similar spongy degeneration may occur in other diseases, although the electron microscopy of Canavan disease is considered to be rather specific (Adachi et al 1972; Adornato et al 1972). Since the basic defect for Canavan disease was not known, the most widely accepted idea for the spongy degeneration was disturbance in the cell ion pump that leads to oedema and sponginess (O'Brien et al 1985).

Adachi and coworkers (1973) have described three clinical variants of the disease: (a) congenital form in which the disease is apparent at birth, or shortly thereafter; (b) an infantile form that is the most common and in which symptoms manifest after the first 6 months of life (this is the form frequently seen among Jewish patients); and (c) a juvenile form in which the disease manifests after the first 5 years of life. Indeed, in our group of patients with Canavan disease the majority have the infantile form, although the life span of these patients varies greatly.

Matalon and colleagues in 1988 discovered increased excretion of *N*-acetylaspartic acid (NAA) and aspartoacylase deficiency in Canavan disease. The initial findings of 3 patients was later extended to include 14 patients (Matalon et al 1989, 1990). *N*-Acetylaspartic aciduria and the deficiency of aspartoacylase in Canavan disease has since been confirmed by several groups (Divry and Mathieu 1989; Echenne et al 1989; Ozand et al 1990; Michelakakis et al 1991). The estimation of NAA levels by gas chromatography–mass spectrometry (GCMS) in the urine from suspected Canavan patients is now routine and the preferred method for the diagnosis of Canavan disease.

N-Acetylaspartic acid, the substrate for aspartoacylase, was discovered in 1956 by Tallan and colleagues (1956), in mammalian brains including humans. Brain is the only organ where biosynthesis of NAA has been demonstrated (Birken and Oldendorf 1989). The concentration of NAA in the human brain is very high, 6–7 μmol/g tissue, with the highest concentration in the cerebral cortex and the lowest in the medulla (Miyake et al 1980, 1981). Recently, using immunohistochemical techniques, NAA has been localized to neurons in rat brain (Moffett et al 1991). The level of NAA is second only to that of glutamic acid among the free amino acid pool of the brain and is higher in concentration than γ-aminobutyric acid (GABA). In spite of its abundance in the brain, the function of NAA is not understood and it has been referred to as an inert metabolite (Jacobson 1957; McIntosh and Cooper 1965; Shigematsu et al 1983). McIntosh and Cooper (1965) suggested that NAA is metabolized slowly in the mature brain. Shigematsu and coworkers (1983) have suggested that NAA is an essential component in a series of reactions required for the conversion of lignoceric acid to cerebronic acid, a component of myelin, and the formation of glutamic acid. High concentrations of NAA-containing peptides in synaptic vesicles and synaptosomes have implicated its role in neurotransmission and production of N-acetylaspartylglutamate (Birken et al 1989). Our preliminary data suggest that NAA in brain may serve as a chemical compartmentation so that aspartate is released only by the action of aspartoacylase. Then the aspartate released may be channelled to form arginine.

Establishing the biochemical defect of the enzyme deficiency in Canavan disease has been a significant step in providing an accurate method for the diagnosis of Canavan disease, identifying carriers who have less than 50% of aspartoacylase activity in cultured fibroblasts, and providing the possibility of prenatal diagnosis.

DIAGNOSIS AND CARRIER TESTING OF CANAVAN DISEASE

Since the initial report of the defect in Canavan disease, we have diagnosed 144 cases through elevated NAA in the urine, blood or brain and/or aspartoacylase deficiency in cultured skin fibroblasts. The specificity of increased NAA excretion in the urine of patients with Canavan disease has been further documented by lack of increased NAA levels in the urine of patients with other leukodystrophies (Matalon et al 1989). The increased level of NAA in urine or brain of Canavan patients is shown in Table 1. Excessive accumulation of NAA in Canavan brain has also been confirmed by ^{1}H-NMR spectroscopy studies (Matalon et al 1990; Grodd et al 1990).

The residual aspartoacylase activity in cultured skin fibroblasts from subjects with Canavan disease, carriers and controls is shown in Table 2. Aspartoacylase activity in the cultured skin fibroblasts from carriers is about half that found in normal fibroblasts. Patients with Canavan disease often have unmeasurable aspartoacylase activity. The aspartoacylase activities from autopsy brain samples from Canavan patients diagnosed by brain biopsy and those from non-Canavan individuals are shown in Table 2. These results document that aspartoacylase is deficient in Canavan brain.

Table 1 Identification of patients with Canavan disease by elevated *N*-acetylaspartic acid (NAA) in urine or brain

Subjects	n	Urine NAA (μmol/mmol creatinine) Mean $\pm$ SD (range)	n	Brain NAA (μmol/mg protein) Range
Canavan	95	1440.5 $\pm$ 873.3 (135.0–3685.9)	11	0.12–0.39
Controls	54	23.5 $\pm$ 16.1 (4.1–59.2)	5	0.00–0.03

Table 2   Aspartoacylase activity in fibroblasts and brain of Canavan disease
patients, carriers and controls

Subject	n	Fibroblasts (mU/mg protein) Mean ± SD (range)	n	Brain (mU/mg protein) Range
Canavan	102	0.005 ± 0.009 (0.000–0.032)	11	0.0–0.05
Control	63	0.240 ± 0.095 0.101–0.493	5	0.12–0.39
Carrier	109	0.070 ± 0.036 0.027–0.133		

PRENATAL DIAGNOSIS OF CANAVAN DISEASE

Prenatal diagnosis of 30 pregnancies at risk for Canavan disease has been performed. Aspartoacylase was assayed in cultured chorionic villus samples (CVS) and amniocytes, and/or N-acetylaspartic acid (NAA) was assayed using stable $(D)_5$ NAA isotope dilution in amniotic fluid (AF). Using the enzyme assay four pregnancies were identified and confirmed to be affected. However, 4 of 24 pregnancies predicted to be normal were abnormal (Matalon et al 1992). The low activity of aspartoacylase in normal cultured CVS and amniocytes makes prenatal diagnosis of Canavan disease by aspartoacylase determination unreliable. It seems that NAA levels in amniotic fluid may be more predictive. Obviously, molecular diagnosis using RFLP or mutation analysis within given families is needed and can be used for more accurate prenatal diagnosis.

PURIFICATION, CHARACTERIZATION AND LOCALIZATION OF ASPARTOACYLASE

Aspartoacylase has been purified to apparent homogeneity with a specific activity of 19 000 nmol of aspartate released/min per mg of protein (Kaul et al 1991a). The native enzyme is a 58-kDa monomer. The purified aspartoacylase activity is enhanced by divalent cations, non-ionic detergents and dithiothreitol. Low levels of dithiothreitol or β-mercaptoethanol are required for enzyme stability. Aspartoacylase has a K_m of 8.5×10^{-4} mol/L and a V_{max} of 43 000 nmol/min per mg of protein. Inhibition of aspartoacylase by glycyl-L-aspartate and amino derivatives of D-aspartic acid suggests that the carbon backbone of the substrate is primarily involved in its interaction with the active site and that a blocked amino group is essential for the catalytic activity of aspartoacylase. Biochemical and immunocytochemical studies revealed that aspartoacylase is localized to white matter, whereas the N-acetylaspartic acid concentration is three-fold higher in grey matter than in white matter (Kaul et al 1991a,b; Johnson et al 1989). Our studies so far indicate that aspartoacylase is conserved across species during evolution and suggest a significant role for aspartoacylase and N-acetylaspartic acid in normal brain biology.

ISOLATION AND CHARACTERIZATION OF BOVINE ASPARTOACYLASE cDNA: AMINO ACID SEQUENCE OF ASPARTOACYLASE

Purified bovine brain aspartoacylase was digested with cyanogen bromide in 70% formic acid (Kaul et al 1983) and four different peptides purified from this digest were sent to Dr K. Williams (Protein Sequencing Facility at Yale University) for determination of the amino acid sequence. Based on the amino acid sequence of the aspartoacylase peptides, oligonucleotides were synthesized using phosphoramidite chemistry (380B DNA synthesizer Applied Biosystems).

Bovine kidney and brain poly(A)$^+$ RNA were reverse-transcribed (RT) for first-strand cDNA synthesis with oligo(dT)$_{18}$ as the primer. The first-strand cDNA was then amplified by the polymerase chain reaction (PCR) with Taq polymerase using CD5 and (dT)$_{18}$ as a set of primers. The amplification product from the oligo(dT)$_{18}$ and CD5 PCR reaction was used as a template with the CD5/CD8 oligo 'cassette' as primers. This procedure resulted in 69-bp cDNA fragments derived originally from kidney and brain poly(A)$^+$ RNA and both had identical restriction maps, suggesting that the two RT–PCR products were derived from a common mRNA species (Figure 1). The fragment was subcloned and the nucleotide sequence was determined. The amino acid sequence predicted from the nucleotide sequence of the 69-bp clone (p2-13) matched the amino acid sequence of the corresponding peptide from purified aspartoacylase.

The cloned 69-bp fragment was used as a probe to screen bovine cDNA libraries. One clone isolated from these libraries was found to have the aspartoacylase-specific coding sequence.

ISOLATION AND CHARACTERIZATION OF BOVINE GENOMIC CLONES

We screened about 10^6 phages each of bovine lymphocyte and placental genomic DNA libraries cloned into the EMBL3 vector (host: LE 392) (Clonetech). The cDNA insert from the bovine clone labelled with ^{32}P was used as a probe and three genomic clones were isolated.

All three phage clones shared a 4-kbp EcoRI band that migrates identically upon digestion with different restriction endonucleases. In addition, one clone had a 6-kbp EcoRI and a 3.4-kbp EcoRI/XhoI band. The 4- and 6-kbp EcoRI bands observed in these clones were similar in size to those observed with Southern blotting analysis of bovine genomic DNA. The restricted fragments from phage DNA inserts were subcloned, and more than 50% of the nucleotide sequence of unique fragments from the three overlapping phage clone inserts have been determined. We have identified six exons so far. The splice junction sequences at the exon/intron boundaries follow the 'gt/ag' rule.

ISOLATION AND CHARACTERIZATION OF HUMAN ASPARTOACYLASE cDNA CLONE

Oligonucleotides based upon the bovine aspartoacylase cDNA sequence were used as primers in the RT–PCR amplification of human aspartoacylase mRNA from human brain and kidney sources. The overall strategy is shown in Figure 2. After initial reverse transcription of poly(A)$^+$ RNA with oligo(dT), the first-strand cDNA was used as template for amplification of aspartoacylase-specific cDNA sequences. A 675-bp cDNA fragment was obtained. The fragment was cloned into a phagemid vector and the nucleotide sequence was determined. Comparison of the nucleotide sequence of human cDNA with the corresponding segment of bovine aspartoacylase cDNA clone revealed a better than 90% homology both at the nucleotide and the amino acid sequence. The amino acid sequence predicted from the cDNA showed a 90–100% homology in different segments of the isolated cDNA. The strong homology seen at the primary structure level corroborates our earlier observation of a highly conserved protein that was based upon studies with aspartoacylase from bovine and human sources (Kaul et al 1991a).

Canavan disease now can be readily diagnosed using the biochemical markers of excess NAA in the urine and deficiency of aspartoacylase. The enzyme determination requires cultured skin fibroblasts. Aspartoacylase activity can also be utilized for carrier detection. Prenatal diagnosis may prove to be more reliable based on quantitation of NAA in the amniotic fluid. The isolation of DNA clones is currently being used for mutation analysis and RFLP determination in patients with Canavan

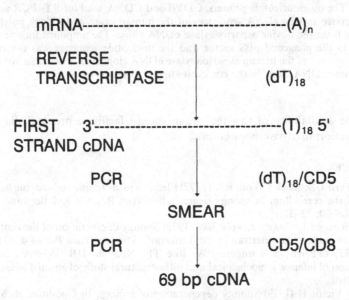

Figure 1 Schematic representation of the strategy for isolation of bovine aspartoacylase cDNA. Bovine brain and kidney mRNA was reverse-transcribed followed by polymerase chain reaction (RT–PCR) amplification of aspartoacylase-specific cDNA sequences. The 69-bp amplification fragment was subcloned in a phagemid pBS vector and the nucleotide sequence of the insert was determined by the dideoxy chain termination method. The primers CD5 and CD8 were based upon the amino acid sequence of the cyanogen bromide peptide CN30 isolated from purified bovine brain aspartoacylase. The amino acid sequence predicted from the 69-bp fragment matched the CN30 peptide sequence. The 69-bp fragment was subsequently used as a probe for isolation of bovine aspartoacylase cDNA clone

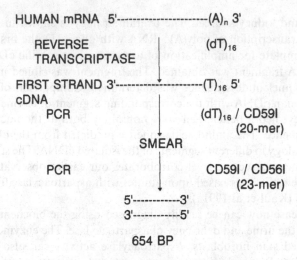

Figure 2 Schematic representation of the strategy for isolation of human aspartoacylase cDNA clone. The overall experimental approach was similar to that described in the legend to Figure 1. The oligonucleotide primers, CD59I and CD56I, used for RT–PCR amplification of aspartoacylase-specific cDNA sequences of the human brain and kidney poly(A)$^+$ RNA$_1$ were derived from the bovine aspartoacylase cDNA clone. The amplified fragment of 654 bp was cloned in the phagemid pBS vector and the nucleotide sequence was determined. The nucleotide sequence of the human aspartoacylase cDNA clone had better than 90% homology with the bovine cDNA clone in the corresponding region

disease. The availability of this technique should facilitate prenatal diagnosis and carrier detection in at-risk populations.

REFERENCES

Adachi M, Torii J, Schneck L, Volk BW (1972) Electron microscopic and enzyme histochemical studies of the cerebellum in spongy degeneration (van Bogaert and Bertrand type). *Acta Neuropathol* **20**: 22–31.

Adachi M, Schneck L, Cazara J, Volk BW (1973) Spongy degeneration of the central nervous system (van Bogaert and Bertrand type; Canavan's Disease). *Hum Pathol* **4**: 331–346.

Adornato BT, O'Brien JS, Lampert PW, Roe TF, Neustein HB (1972) Cerebral spongy degeneration of infancy: a biochemical and ultrastructural study of affected twins. *Neurology* **22**: 202–210.

Banker BQ, Victor H (1979) Spongy degeneration of infancy. In Goodman R, Motulsky A, eds. *Genetic Diseases Among Ashkenazi Jews*. New York: Raven Press, 201–217.

Banker BQ, Robertson JJ, Victor M (1964) Spongy degeneration of the central nervous system in infancy. *Neurology* **14**: 981–1001.

Birken DL, Oldendorf WH (1989) *N*-Acetyl-L-aspartic acid: A literature review of a compound prominent in ^{1}H-NMR spectroscopic studies of brain. *Neurosci Behav Rev* **13**: 23–31.

van Bogaert L, Bertrand I (1949) Sur une idiotie familiale avec degerescence songlieuse de neuraxe (note preliminaire). *Acta Neurol Belg* **49**: 572–587.

van Bogaert L, Bertrand I (1967) *Spongy Degeneration of Brain in Infancy*. Amsterdam: North Holland, 3–132.

Buchanan DS, Davis RL (1965) Spongy degeneration of the nervous system: a report of 4 cases with a review of the literature. *Neurology* **15**: 207–222.

Canavan MM (1931) Schilder's encephalitis periaxialis diffusa. *Arch Neurol Psychiatr* **25**: 299–308.

Divry P, Mathieu M (1989) Aspartoacylase deficiency and N-acetylasparticaciduria in patients with Canavan disease. *Am J Med Genet* **32**: 550.

Echenne B, Divry P, Viamey-Liaud C (1989) Spongy degeneration of the neuraxis (Canavan–Van Bogaert disease) and N-acetylaspartic aciduria. *Neuropediatrics* **20**: 179–181.

Gambetti P, Mellman WJ, Gonatas NK (1969) Familial spongy degeneration of the central nervous system (van Bogaert–Bertrand disease). *Acta Neuropathol* **12**: 103–115.

Globus JH, Strauss I (1928) Progressive degenerative subcortical encephalopathy (Schilder's disease). *Arch Neurol Psychiatr* **20**: 1190–1228.

Grodd W, Kragelh-Mann I, Peterson D, Treftz FK, Harzer K (1990) In vivo assessment of N-acetylaspartate in brain in spongy degeneration (Canavan's disease) by proton spectroscopy. *Lancet* **336**: 437–438.

Hagenfeldt L, Bollgren I, Venizelos N (1987) N-Acetylaspartic aciduria due to aspartoacylase deficiency – a new etiology of childhood leukodystrophy. *J Inher Metab Dis* **10**: 135–141.

Jacobson KB (1957) Studies on the role of N-acetylaspartic acid on mammalian brain. *J Gen Physiol* **43**: 323–333.

Johnson AB, Kaul RK, Casanova J, Matalon R (1989) Aspartoacylase, the deficiency enzyme in spongy degeneration (Canavan disease) is a myelin-associated enzyme. *J Neuropathol Exp Neurol* **48**: 349.

Kaul RK, Murphy SNP, Reddy AG, Steck TL, Kohler H (1983) Amino acid sequence of the N-terminal 201 residues of human erythrocyte membrane band 3. *J Biol Chem* **258**: 7981–7990.

Kaul RK, Casanova J, Johnson A, Tang P, Matalon R (1991a) Purification, characterization and localization of aspartoacylase from bovine brain. *J Neurochem* **56**: 129–135.

Kaul R, Michals K, Casanova J, Matalon R (1991b) The role of N-acetylaspartic acid in brain metabolism and the pathogenesis in Canavan disease. *Int Pediatr* **6**: 40–43.

Kvittingen EA, Guldal G, Borsting S, Skalpe IO, Stokke O, Jellum E (1986) N-Acetylaspartic aciduria in a child with a progressive cerebral atrophy. *Clin Chim Acta* **158**: 217–227.

Matalon R, Michals K, Sebasta D, Deanching M, Gashkoff P, Casanova J (1988) Aspartoacylase deficiency and N-acetylaspartic aciduria in patients with Canavan disease. *Am J Med Genet* **29**: 463–471.

Matalon R, Kaul RK, Casanova J et al (1989) Aspartoacylase deficiency: the enzyme defect in Canavan disease. *J Inher Metab Dis* **12**: 329–331.

Matalon R, Michals K, Kaul R, Mafee M (1990) Spongy degeneration of the brain, Canavan disease. *Int Pediatr* **5**: 121–124.

Matalon R, Michals K, Gashkoff P, Kaul R (1992) Prenatal diagnosis of Canavan disease. *J Inher Metab Dis* **15**: 392–394.

McIntosh JM, Cooper JR (1965) Studies on the function of N-acetylaspartic acid in the brain. *J Neurochem* **12**: 825–835.

Menkes JH (1985) *Textbook of Child Neurology*, 3rd edn. Philadelphia: Lea & Febiger.

Michelakakis H, Giouroukos S, Divry P, Katsarou E, Rolland MO, Skardoutsow A (1991) Canavan disease: findings in four new cases. *J Inher Metab Dis* **14**: 267–268.

Miyake M, Kakimoto Y (1981) Developmental changes of N-acetyl-L-aspartic acid, N-acetyl-alpha-aspartylglutamic acid and beta-citryl-L-glutamic acid in different brain regions and spinal cords of rat and guinea pig. *J Neurochem* **37**: 1064–1067.

Miyake M, Kakimoto Y, Sorimachi M (1980) A gas chromatographic method for the determination of N-acetyl-L-aspartic acid, N-acetyl-alpha-aspartylglutamic acid and beta-citryl-L-glutamic acid and their distributions in the brain and other organs of various species of animals. *J Neurochem* **36**: 804–810.

Moffett JR, Namboodiri MAA, Cangro CB, Neale JH (1991) Immunohistochemical localization of N-acetylaspartate in rat brain. *Neuro Report* **2**: 131–134.

O'Brien DP, Zachary JF (1985) Clinical features of spongy degeneration of the central nervous system in two Labrador retriever littermates. *J Am Vet Med Assoc* **186**: 1207–1210.

Ozand PT, Gascon G, Dhalla M (1990) Aspartoacylase deficiency and Canavan disease in Saudi-Arabia. *Am J Med Genet* **35**: 266–268.

Rushton AR, Shaywitz BA, Dumen CC, Geehr RB, Maneulidis EE (1981) Computerized tomography in the diagnosis of Canavan's disease. *Ann Neurol* **10**: 57–60.

Sacks O, Brown WJ, Aguilar MJ (1965) Spongy degeneration of white matter. Canavan's sclerosis. *Neurology* **15**: 165–171.

Shigematsu H, Okamura N, Shimeno H, Kishimoto Y, Kan L, Fenselau C (1983) Purification and characterization of the heat stable factors essential for conversion of lignoceric acid to cerebronic acid and glutamic acid: identification of *N*-acetyl-L-aspartic acid. *J Neurochem* **40**: 814–820.

Tallan HH, Moore S, Stein WH (1956) *N*-acetyl-L-aspartic acid in brain. *J Biol Chem* **219**: 257–264.

Ungar M, Goodman RM (1983) Spongy degeneration of the brain in Israel: A retrospective study. *Clin Genet* **23**: 23–29.

J. Inher. Metab. Dis. 16 (1993) 753–761
© SSIEM and Kluwer Academic Publishers. Printed in the Netherlands

L-2-Hydroxyglutaric Acidaemia: Clinical and Biochemical Findings in 12 Patients and Preliminary Report on L-2-Hydroxyacid Dehydrogenase

P. G. Barth[1,2], G. F. Hoffmann[3], J. Jaeken[4], R. J. A. Wanders[1],
M. Duran[5], G. A. Jansen[1], C. Jakobs[6], W. Lehnert[7], F. Hanefeld[8],
J. Valk[9], R. B. H. Schutgens[1], F. K. Trefz[3], H.-P. Hartung[10],
N. A. Chamoles[11], Z. Sfaello[11], U. Caruso[12]

Departments of [1]Pediatrics and [2]Neurology, University Hospital Amsterdam, Meibergdreef 9, NL-1105 AZ, Amsterdam; The Netherlands; [3]Department of Pediatrics and Metabolic Diseases, Children's Hospital, University of Heidelberg, Germany; [4]Department of Pediatrics, University Hospital, Leuven, Belgium; [5]Division of Metabolic Diseases, University Children's Hospital, Utrecht, The Netherlands; [6]Department of Pediatrics, Free University Hospital, Amsterdam, The Netherlands; [7]Department of Pediatrics, Children's Hospital, University of Freiburg, Germany; [8]Department of Neuropediatrics, Children's Hospital, University of Göttingen, Germany; [9]Department of Neuroradiology, Free University Hospital, Amsterdam, The Netherlands; [10]Department of Neurology, University of Würzburg, Germany; [11]Laboratory of Neurochemistry, Uriarte 2383, 1425 Buenos Aires, Argentina; [12]Department of Pediatrics I, University of Genova, Italy

Summary: L-2-Hydroxyglutaric acidaemia represents a newly defined inborn error of metabolism, with increased levels of L-2-hydroxyglutaric acid in urine, plasma and cerebrospinal fluid. The concentration in cerebrospinal fluid is higher than in plasma. The other consistent biochemical finding is an increase of lysine in blood and cerebrospinal fluid, but lysine loading does not increase L-2-hydroxyglutaric acid concentration in plasma. This autosomal recessively inherited disease is expressed as progressive ataxia, mental deficiency with subcortical leukoencephalopathy and cerebellar atrophy on magnetic resonance imaging. Since these features were described in 8 patients by Barth and co-workers in 1992, 4 more patients with similar findings have been diagnosed and added to the present series. L-2-Hydroxyglutaric acid is found in only trace amounts on routine gas chromatographic screening in normal persons, and its origin, its fate and even its relevance to normal metabolism are unknown. Therefore its catabolism was studied in normal liver. Incubation of rat liver with L-2-hydroxyglutaric acid did not produce H_2O_2, which excluded (peroxisomal) L-2-hydroxyacid oxidase as the main route of catabolism. However, L-2-hydroxyglutaric acid is rapidly dehydrogenated if NAD^+ is added as a co-factor to the standard reaction medium. This could also be demonstrated in human liver. The preliminary evidence for this enzyme activity in rats and humans, L-2-hydroxyglutaric acid dehydrogenase, is given. Further investigations are required to clarify the possible relevance to the metabolic defect in L-2-hydroxyglutaric acidaemia.

Since the description of the first case of L-2-hydroxyglutaric aciduria (Duran et al 1980) in a 5-year-old boy with psychomotor retardation and dystrophy, an abnormality of white matter suggestive of leukodystrophy has been reported in other patients (Jaeken et al 1988; Hoffmann et al 1990). A recent report (Barth et al 1992) was presented based on 8 patients from various European countries, including the patients previously reported. Autosomal recessive inheritance was strongly suggested by the family histories. Similar neurological abnormalities were present in each case, with a progressive course of mental deficiency, ataxia and other motor system deficiencies. Increased L-2-hydroxyglutaric acid was found in urine, plasma and CSF of all patients investigated. Computerized tomography (CT), and especially magnetic resonance imaging (MRI) showed a consistent pattern of subcortical leukoencephalopathy. Furthermore, cerebellar atrophy and signal changes in several nuclei were found, together representing a pattern not found in other known neurodegenerative disorders.

In this communication four more patients are included in the series, giving added evidence for the genetic origin, the specific pattern of neurological affection involved and the accompanying abnormalities on MRI. Comparison is made to L-glutaric aciduria type I (McKusick 23167).

In order to find a metabolic role for L-2-hydroxyglutaric acid, its metabolic fate in homogenates of normal rat and human liver was studied by enzymatic methods and the findings – a newly discovered type of L-2-hydroxyglutaric acid dehydrogenase activity – are reported.

METHODS

Organic acids in body fluids were assayed by GC–MS. Identification of L-2-hydroxyglutaric acid and determination of its absolute configuration as the L-stereoisomer were performed as previously described (Duran et al 1980). Amino acids were quantified by automatic amino acid analysis. Pipecolic acid was determined as previously described (Kok et al 1987).

Fibroblasts from patients and controls were incubated with $12\,\mu$Ci $[1-^{14}C]$-2-oxoglutaric acid for 48 h. Thereafter the medium and the cells were combined, organic acids were separated by liquid partition chromatography as previously described (Sweetman 1974) and the radioactivity in 2-hydroxyglutaric acid was determined. The amount of 2-oxoglutaric acid metabolized via the citric acid cycle was estimated by determining the production of $^{14}CO_2$.

The activities of L-2-hydroxyglutaric acid oxidase and L-2-hydroxyglutaric acid dehydrogenase in homogenates of rat liver as well as human livers were measured according to the following procedures. For L-2-hydroxyglutaric acid oxidase activity measurements, the reaction medium contained the following standard components: 100 mmol/L Tris-HCl (pH 8.8), 0.05% (w/v) Triton X-100, 6 mmol/L 2,4,6-tribromohydroxybenzoate and 1 mmol/L aminoantipyrine. Reactions were initiated by adding

10 mmol/L L-2-hydroxyglutarate (sodium salt) and the absorbance was measured at 510 nm as previously described (Wanders et al 1989). For L-2-hydroxyglutarate dehydrogenase activity measurements, the same reaction medium was used except that 5 mmol/L NAD$^+$ was added. Reactions were started by adding 10 mmol/L L-hydroxyglutarate and the absorbance at 340 nm was followed using a COBAS-BIO centrifugal analyser (Hoffman-LaRoche, Basle, Switzerland).

PATIENTS

Clinical findings: These are summarized in Table 1. As a rule no abnormalities were noted in the first year of life. Thereafter, delay in unsupported walking (A1, B1, D1, E2), abnormal gait (C1), speech delay (A2), or severe febrile seizures (D5) were the presenting symptoms in early childhood in 7 patients. In 4 patients no abnormalities were noted until learning disabilities in the first school years drew attention to their handicap. In one patient (E1) cerebellar signs at age 10 years were the first symptom. Seizures, either febrile or grand mal or both, were present in 6 patients. The main motor handicap was cerebellar dysfunction (dysarthria, truncal ataxia, dysmetria and gait ataxia are reported individually and in variable combinations). Cerebellar symptoms were indeed the most common abnormality, after mental deficiency. These were present in all patients over 15 years of age.

There are strong indications for autosomal recessive inheritance, since 4 of 7 families had more than one affected child, and in two families the parents were consanguineous (D and G). Males and females are similarly affected. The ratio between affected and unaffected sibs is 12/12. Further reporting will enlarge the series and may give the true proportions between males and females, as well as the proportion between affected and unaffected sibs. The finding that 4 of the 7 families had more than one affected child may in part be explained by increased demand for expert diagnostic help in the case of families with more than one affected child, and

Table 1   Clinical profiles of affected patients

Patient no.[a]	Sex	Age[b] (years)	Mental deficiency	Cerebellar symptoms	Extrapyramidal symptoms	Pyramidal symptoms
A1	M	19	+	+	+	−
A2	M	13	+	−	−	−
B1	F	20	+	+	+	−
C1	F	16	+	+	−	+
D1	F	19	+	+	−	−
D2	F	15	+	+	+	−
E1	M	39	+	+	+	+
E2	M	33	+	+	−	−
F1	F	16	+	+	−	−
F2	F	16	+	+	−	−
F3	F	11	+	+	−	−
G	F	12	+	+	−	−

[a]Affected sibs are indicated by same letter
[b]Age at last follow-up

J. Inher. Metab. Dis. 16 (1993)

this in turn may explain the 1/1 ratio of affected/unaffected sibs, rather than the 1/4 expected on the basis of autosomal recessive inheritance. The ethnic origins of the affected families were in different regions: Morocco, Turkey, Greece, Germany, Italy and Latin America.

Neuroimaging studies: The results of neuroimaging have been published previously for the patients A1 through E2 (Barth et al 1992). Briefly, the most specific abnormality was loss of subcortical white matter. This was demonstrable by CT, but could be demonstrated more convincingly by MRI, where loss of signal in T_1-weighted sections and increased signal on T_2-weighted sections was found in the subcortical regions. The internal and external capsules were moderately involved. Periventricular white matter was much less or not involved. The caudate nuclei were atrophic; the putamen showed signal changes; the pallidum, thalamic nuclei, mesencephalon and lower brainstem seemed uninvolved. The cerebellum was affected by folial atrophy, mainly affecting the vermis, and signal changes in the dentate nuclei. Two new MRI series not previously described were added to this patient series (patients F1 and G1). Review of their MRI disclosed the same pattern as previously described, except that in the case of patient G1 there was more extensive involvement of the deep frontal white matter in addition to the typical findings described above. A typical example is given in Figure 1.

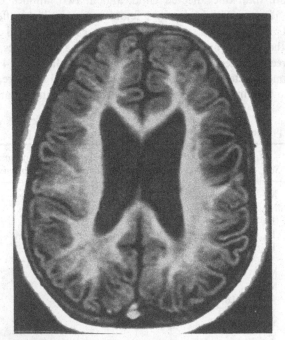

Figure 1 T_1-weighted axial MR section from patient A1. Subcortical white matter is severely deficient, shown by the nearly empty appearance of the gyral cores. There is moderate signal loss in the central white matter

RESULTS

Results of determinations of L-2-hydroxyglutaric acid in plasma, CSF and urine, and of lysine in plasma and CSF are given in Table 2. Elevations were found in each patient. The results of paired samples from CSF and plasma showed that the concentration of L-2-hydroxyglutaric acid in the CSF was higher than in plasma in all cases where paired samples were available ($n = 6$). This is represented in Figure 2, where paired plasma and CSF concentrations are plotted against each other. Lysine was elevated in all samples of plasma and CSF examined. Plotting of the concentrations of plasma and CSF against each other shows that plasma levels were higher than CSF except in one patient (Figure 3).

Table 2   Concentration of metabolites in body fluid

Body fluid	Patients		Controls	
	2OHglu ac.[a]	Lysine	2OHglu ac.[a]	Lysine
Plasma (μmol/L)				
Range	7–84	70–380	ND[c]	120–230
Mean[b]	31 (10)	279 (8)		
CSF (μmol/L)				
Range	23–474	66–89	ND	10–25
Mean	122 (6)	79 (6)		
Urine (mmol/g creatinine)				
Range	2–38	0.10–0.37	< 0.46	0.05–0.40
Mean	16 (12)	0.24 (4)		

[a]2OHglu ac. = L-2-hydroxyglutaric acid
[b]Number of patients in parentheses
[c]ND = not detectable

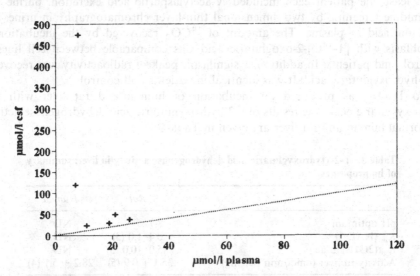

Figure 2 L-2-Hydroxyglutaric acid concentrations in plasma and CSF samples drawn at the same time plotted against each other

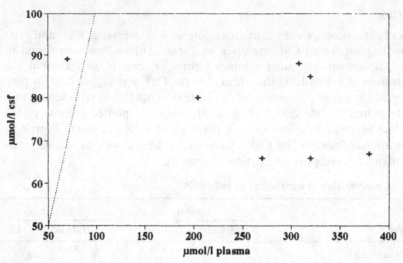

Figure 3 Lysine concentrations in plasma and CSF samples drawn at the same time plotted against each other

Pipecolic acid was determined in plasma and in CSF from patients E1 and E2 with normal results. No excesses of L-2-hydroxyacids other than L-2-hydroxyglutaric acid were found by the procedures employed. 2-Oxoglutaric acid was not elevated in any of the samples of plasma or CSF investigated. Long-chain and very long-chain fatty acids (C_{16} to C_{26}) were examined by GC–MS in patients A1 and A2, with normal results. Furthermore, miscellaneous investigations with normal results in at least one patient each included N-acetylaspartic acid excretion, purine and pyrimidine screening by two-dimensional thin-layer chromatography in urine, and phytanic acid in plasma. The amount of $^{14}CO_2$ recovered by the incubation of fibroblasts with [1-^{14}C]-2-oxoglutaric acid was comparable between cell lines of controls and patients. In addition, a significant peak of radioactivity was recovered in 2-hydroxyglutaric acid. It was identical in patients and controls.

No H_2O_2 was produced on incubation of human and rat liver with L-2-hydroxyglutaric acid. The results of L-2-hydroxyglutaric acid dehydrogenase activity in normal human and rat liver are given in Table 3.

Table 3 L-2-Hydroxyglutaric acid dehydrogenase activity in liver: summary of its properties

	Rat	Human
pH optimum	8.7	ND[a]
K_m (NAD$^+$)	0.4–1.0 (3)	ND
K_m (2OH-glu ac.)[b]	5.0–10.0 (3)	ND
Activity in liver (nmol/min per mg)	35.1 ± 0.9 (5)	28.2 ± 5.1 (4)

[a]ND = not determined
[b]2OHglu ac. = L-2-hydroxyglutaric acid

DISCUSSION

The disease pattern described above is fairly specific from the viewpoint of neurological and MRI findings. It should be pointed out that no periods of metabolic acidosis and no acute deteriorations were found in any of the patients, but rather a chronic, slowly progressive disorder with insidious onset after infancy. A related compound, glutaric acid, is specifically elevated in glutaric acidaemia type I, which results from a defect of glutaryl-CoA dehydrogenase (EC 1.3.99.7) in the degradation of lysine and tryptophan.

The history of patients with glutaric acidaemia type I is very different from that of patients with L-2-hydroxyglutaric acidaemia. In the former, the usual picture is one of early-onset dystonic cerebral palsy, in most cases with an acute onset in infancy (Haworth et al 1991; Hoffmann et al 1991). Diminished serum free L-carnitine and intermittent dicarboxylic aciduria point to secondary involvement of mitochondrial fatty acid metabolism. In the case of L-2-hydroxyglutaric acidaemia no acute encephalopathic episodes, no abnormalities of carnitine metabolism or excess of dicarboxylic acids have been found in any of the patients investigated. L-2-Hydroxyglutaric acid is higher in CSF than in plasma in all 7 cases thus far investigated. This suggests an endogenous origin of the compound, although a bacterial origin has been described (Reeves and Ajl 1962). The increased CSF-to-plasma ratio may be due to a specific role of L-2-hydroxyglutaric acid in cerebral metabolism.

The stereotyped MRI pattern is a subcortical encephalopathy, whereas myelin loss is mainly periventricular in all the leukodystrophies known. For the pattern of demyelination, comparison may be drawn with Canavan's spongiform encephalopathy (aspartoacylase deficiency), where myelin breakdown also is predominantly subcortical rather than periventricular. Together with the other findings, especially the cerebellar atrophy and the signal changes in several nuclei that are important in the organization of voluntary movement, such as the putamen and the caudate and dentate nuclei, the pattern is highly specific. Therefore, it should easily be recognized by neuroradiologists while investigating children or adults for mental deficiency and progressive ataxia. This finding should prompt GC–MS investigation of urine, plasma and CSF, when this has not been done previously.

So far there is no definite clue to the metabolic block involved. A number of loading and fasting studies have been conducted without conclusive results (Barth et al 1992). The consistent elevation of lysine in the patients drew our attention to lysine catabolism. We previously reported that L-lysine loading (75 mg/kg body weight) in two of the patients did not result in any change of the basal plasma level of L-2-hydroxyglutaric acid, even though the absorption of lysine was quite efficient (Barth et al 1992). Since the principal catabolic pathway for L-lysine in the brain has been reported to be the pipecolic acid pathway, rather than the saccharopine pathway (Chang 1976) the levels of pipecolic acid in CSF and in plasma were determined in two patients. The results were normal. Therefore, there is little evidence for a primary block in lysine catabolism to explain the findings. Earlier, Jaeken and colleagues (1988) proposed that the raised lysine could originate from a possible interference of an excess of L-2-hydroxyglutaric acid with lysine breakdown through competitive inhibition, possibly with 2-oxoglutarate at the level of saccharopine dehydrogenase.

Two approaches were tried to gain information on the metabolic pathways related to L-2-hydroxyglutaric acid. First, we tried to obtain information on the formation of 2-hydroxyglutaric acid by studying the turnover of radiolabelled 2-oxoglutaric acid in fibroblasts of patients and controls. By using $[1-^{14}C]$-2-oxoglutaric acid, labelled at the first carbon, the label of the compound should have been lost if it was metabolized via the citric acid cycle. The amount of 2-oxoglutaric acid metabolized via the citric acid cycle was estimated by determining the production of $^{14}CO_2$ and was comparable between cell lines of controls and patients. In addition a significant peak of radioactivity was recovered in 2-hydroxyglutaric acid. It was identical in patients and controls. This implied that 2-hydroxyglutaric acid was formed from 2-oxoglutaric acid in human fibroblasts. However, there was no apparent difference between patients and controls. As it was not possible to differentiate between the formation of the two isomers of D- and L-2-hydroxyglutaric acid by this approach, it cannot be excluded that the assay tested the formation of D-2-hydroxyglutaric acid and not L-2-hydroxyglutaric acid.

We then tested the metabolic fate of L-2-hydroxyglutaric acid in tissues. First, we investigated the possibility that L-2-hydroxyglutaric acid reacts with an oxidase, which uses molecular oxygen as a substrate, subsequently yielding hydrogen peroxide (H_2O_2). In the past several L-2-hydroxyacid oxidases have been identified, including L-2-hydroxyacid oxidase A and B in liver and kidney, respectively, from rat and other species (Nakano et al 1968; Yokota et al 1985). These experiments, in which the homovanillic acid/peroxidase system was used to measure the production of hydrogen peroxide (Wanders et al 1989), revealed that L-2-hydroxyglutaric acid is not degraded in an oxidase type reaction, which excluded a significant role for peroxisomal L-2-hydroxyacid oxidases. Subsequently, we investigated whether L-2-hydroxyglutaric acid is degraded in a dehydrogenase type reaction. These studies showed that L-2-hydroxyglutaric acid is indeed rapidly dehydrogenated if NAD^+ is added as a cofactor to the standard reaction medium. The properties of this enzyme activity as listed in Table 3 show that the enzyme as present in rat liver has a relatively high pH-optimum (8.7) and a low affinity for L-2-hydroxyglutaric acid. Importantly, the enzyme activity was found to be about equal in human liver specimens. In the past the occurrence of L-2-hydroxyglutaric acid as an intermediate in rat tissues was reported by Weil-Malherbe (1937), but the significance of this metabolite for man was not raised before the discovery of L-2-hydroxyglutaric acidaemia. Subsequent studies in patients will be required to reveal whether the accumulation of L-2-hydroxyglutaric acid is indeed due to a deficiency of this newly identified enzyme activity. Such studies are underway.

REFERENCES

Barth PG, Hoffmann GF, Jaekenn JJ et al (1992) L-2-Hydrogylutaric acidemia: A novel inherited neurometabolic disease. *Ann Neurol* **32**: 66–71.

Chang Y-F (1976) Pipecolic acid pathway: the major lysine metabolic route in the rat brain. *Biochem Biophys Res Commun* **69**: 174–180.

Duran M, Kamerling JP, Bakker HD, van Gennip AH, Wadman SK (1980) L-2-Hydroxyglutaric aciduria: An inborn error of metabolism? *J Inher Metab Dis* 3: 109–112.

Jaeken J, Willekens H, Corbeel L (1988) Leukodystrophy associated with hyperlysinorrhachia and 2-hydroxyglutaric aciduria. *Pediatr Res* 24: 266 (abstract).

Haworth JC, Booth FA, Chudley AE et al (1991) Phenotypic variability in glutaric aciduria type I: Report of fourteen cases in five Canadian Indian kindreds. *J Pediatr* 118: 52–58.

Hoffmann G, Voss W, Hunneman DH et al (1990) L-2-Hydroxyglutarazidurie: Eine neue Enzephalopathie mit leukodystrophen Veränderungen. In Hanefeld, F, Rating, D, Christen H-J, eds. *Aktuelle Neuropädiatrie 1989.* New York: Springer-Verlag, 139–142.

Hoffmann GF, Trefz FK, Barth PG et al (1991) Glutaryl-coenzyme A dehydrogenase deficiency: A distinct encephalopathy. *Pediatrics* 88: 1194–1203.

Kok R, Kaster M, de Jong L, Poll-Thé B, Saundubray J-M, Jakobs C (1987) Stable isotope dilution analysis of pipecolic acid in CSF, plasma, urine and amniotic fluid using electron capture negative ion mass fragmentography. *Clin Chim Acta* 168: 143–152.

Nakano M, Ushijima Y, Saga M, Tsutsumi Y, Asami H (1968) Aliphatic L-α-hydroxyacid oxidase from rat livers purification and properties. *Biochim Biophys Acta* 167: 9–22.

Reeves HC, Ajl SJ (1962) Alpha-hydroxyglutaric acid synthetase. *J Bacteriol* 84: 186–187.

Sweetman L (1974) Liquid partition chromatography and GC/MS in identification of acid metabolites of aminoacids. In Nyhan WL, ed. *Inheritable Disorders of Aminoacid Metabolism.* New York: Wiley, 730–751.

Wanders RJA, Romeyn GJ, Schutgens RBH, Tager JM (1989) L-Pipecolate oxidase: A distinct peroxisomal enzyme in man. *Biochem Biophys Res Commun* 164: 550–555.

Weil-Malherbe H (1937) The oxidation of 1(−)α-hydroxyglutaric acid in animal tissues. *J Biochem* 31: 2080–2094.

Yokota S, Ichikawa K, Hashimoto T (1985) Light and electron microscopic localization of L-alpha-hydroxyacid oxidase in rat kidney revealed by immunocytochemical techniques. *Histochemistry* 82: 25–32.

J. Inher. Metab. Dis. 16 (1993) 762–770
© SSIEM and Kluwer Academic Publishers. Printed in the Netherlands

Biochemical Pathogenesis of Subacute Combined Degeneration of the Spinal Cord and Brain

R. SURTEES

Institute of Child Health, London, UK

Summary: In humans, subacute combined degeneration of the spinal cord and brain, a primary demyelinating disease, is caused by cobalamin or methyltetrahydrofolate deficiency. Experimental studies into its pathogenesis suggest that dysfunction of the methyl-transfer pathway may be the cause. Compelling evidence for this comes from the study of inborn errors of cobalamin metabolism where deficiency of methylcobalamin, but not deoxyadenosylcobalamin, is associated with demyelination. Recent studies have focused upon inborn errors of the methyl-transfer pathway. Cerebrospinal fluid concentrations of metabolites of the methyl-transfer pathway have been measured in humans with sequential errors of the pathway and correlated with demyelination demonstrated on magnetic resonance imaging of the brain. This has provided new data suggesting that deficiency of S-adenosylmethionine is critical to the development of demyelination in cobalamin deficiency.

Lichtheim (1887) first recognized that progressive myelopathy may accompany pernicious anaemia. Demyelinating lesions were found in the posterior and lateral columns of the spinal cord, and often the anterior columns. This combination of tract degeneration led Russell and colleagues (1900) to suggest the term 'subacute combined degeneration' in the first full clinical and neuropathological description of the disease. The characteristic microscopic appearance was limited to the white matter of the degenerating tracts and showed vacuolation of the myelin sheath. Billings (1902) stressed the constant association between subacute combined degeneration and pernicious anaemia.

Neuropathological involvement of the brain in well-documented pernicious anaemia was recognized by Woltman (1918) but not fully characterized until 1944 (Adams and Kubik). The essential brain pathology is similar to the tract degenerations seen in the spinal cord and medulla, and consists of a diffuse but uneven demyelination in the cerebral white matter occurring mainly around blood vessels. As with the spinal cord lesions, nerve degeneration, when it occurs, is secondary to the destruction of myelin.

AETIOLOGY

Cobalamin deficiency

Success in the use of raw liver to treat pernicious anaemia resulted in the purification of a crystalline anti-pernicious anaemia factor. This factor was vitamin B_{12} or cobalamin. Ungley (1949) showed that administration of cobalamin improved the symptoms and signs of subacute combined degeneration. Since then subacute combined degeneration has been described in patients with various acquired and inherited forms of cobalamin malabsorption (Pallis and Lewis 1974) and in patients with inborn errors of intracellular cobalamin metabolism (Dayan and Ramsey 1974).

The anaesthetic and analgesic gas nitrous oxide inactivates cobalamin by oxidizing the cobalt atom. Abuse of this substance over a prolonged period of time causes a myeloneuropathy very similar to subacute combined degeneration (Layzer et al 1978). This myeloneuropathy can be reversed by stopping exposure to nitrous oxide and administration of cobalamin, suggesting that inactivation of cobalamin also causes subacute combined degeneration.

Methyltetrahydrofolate deficiency

More recently it has been suggested that folate deficiency, without cobalamin deficiency, can cause subacute combined degeneration (Lever et al 1986). However, folic acid administration to patients with pernicious anaemia may precipitate or exacerbate subacute combined degeneration (Israëls and Wilkinson 1949). This effect has been regarded as evidence against the view that folate deficiency may cause subacute combined degeneration. However, folic acid is not a natural folate but instead a stable oxidized form that may compete with the transport of natural reduced tetrahydrofolates into the nervous system (Levitt et al 1971). This would result in central nervous system depletion of tetrahydrofolate because folic acid is reduced by dihydrofolate reductase at a very slow rate and has even been shown to inhibit this enzyme *in vivo* (Blackley 1969; Auletta et al 1974).

Subacute combined degeneration of the cord and brain has been pathologically demonstrated in two children with an inborn error of metabolism causing intracellular methyltetrahydrofolate deficiency (Clayton et al 1986; Beckman et al 1987). Acquired folate deficiency in inborn errors of tetrahydrobiopterin metabolism is also associated with a myeloneuropathy highly suggestive of subacute combined degeneration (Smith et al 1985).

There now seems little doubt that methyltetrahydrofolate deficiency, due to either nutritional deficiency, malabsorption or inborn errors of metabolism, can cause subacute combined degeneration.

PATHOGENESIS

Introduction and biochemistry

The only direct link between cobalamin and folate metabolism is the one-carbon transfer pathway that is shown schematically in Figure 1. One-carbon units enter this pathway from the cytoplasmic pool, mainly derived from serine, and bind to

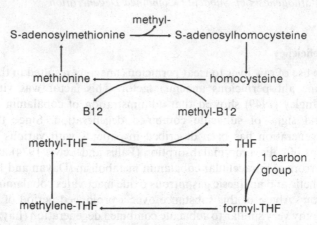

Figure 1 The single-carbon transfer pathway: THF is tetrahydrofolate; B12 is cobalamin. Further details in the text

tetrahydrofolate to form 5,10-methylenetetrahydrofolate. 5,10-Methylenetetrahydrofolate is reduced by 5,10-methylenetetrahydrofolate reductase to give 5-methyltetrahydrofolate. 5-Methyltetrahydrofolate acts as the methyl-donor for the conversion of homocysteine to methionine by methylcobalamin–methionine synthase. Methionine is then activated by methionine adenosyltransferase to form S-adenosylmethionine. Whilst each substituent of the sulphonium atom of S-adenosylmethionine may participate in a transfer reaction, in mammals only the methyl and aminopropyl groups are transferred. In the brain S-adenosylmethionine is the mammalian 'universal' methyl-group donor, acting in a wide variety of biological methylations that modify proteins, nucleic acids, fatty acids, phospholipids and polysaccharides and is necessary for the inactivation of biogenic amines (Cantoni 1975; Zappia et al 1979). In addition, S-adenosylmethionine is the aminopropyl-group donor for the biosynthesis of polyamines and during this series of reactions methiobutyrate is formed, which may be converted to methionine or formate (Tabor and Tabor 1976).

Following biological methylation, S-adenosylhomocysteine is formed. S-Adenosylhomocysteine is a potent inhibitor of all methyltransferases and the ratio of the concentrations of S-adenosylmethionine and S-adenosylhomocysteine (called the methylation ratio) controls the rate of biological methylation (Hoffman et al 1980; Schatz et al 1981). S-Adenosylhomocysteine is catabolized to homocysteine and adenosine by S-adenosylhomocysteine hydrolase. The kinetics of S-adenosylhomocysteine hydrolase favour the formation of S-adenosylhomocysteine, but the efficient removal of adenosine by adenosine deaminase and the remethylation of homocysteine to methionine by methionine synthase or the condensation of homocysteine and serine to form cystathionine by cystathionine synthase drives the reaction towards homocysteine.

Experimental studies

Cobalamin deficiency causing neurological disease ('pine') has been reported in sheep grazing in cobalt-deficient pastures. Monkeys and fruit bats fed a washed vegetarian diet containing no cobalamin develop the neurological disease. In each species demyelination of the spinal cord has been documented.

Inactivation of cobalamin by exposure to nitrous oxide produces neurological disease in monkeys, fruit bats and pigs (Scott et al 1981; van der Westerhuizen et al 1982; Weir et al 1988). In each species the histological lesion is identical to subacute combined degeneration and, in each, administration of methionine has been shown to prevent or ameliorate the neurological disease. Exposure of rats and other species to nitrous oxide does not induce demyelination.

Other studies have examined the effect of disturbances of single-carbon metabolism at sites other than methionine synthase. Administration of cycloleucine (an inhibitor of methionine adenosyltransferase to mice, rats or chicks causes demyelination of the spinal cord identical to subacute combined degeneration (Lee et al 1992; Ramsey and Fisher 1978; Small et al 1981). Cultured mouse cerebral cells do not form compact myelin in the presence of sinefungin, a specific inhibitor of protein methyltransferase I (Amur et al 1986).

These studies have suggested that dysfunction of the methyl-transfer pathway may be the cause of subacute combined degeneration. The inability to form compact myelin in the presence of a protein methylase I inhibitor *in vitro* suggests that the failure of methylation of arginine$_{107}$-myelin basic protein might be an important event. Protein methylase I is specific for the methylation of arginine$_{107}$-myelin basic protein (Ghosh et al 1990).

Biochemical studies of the brain in the above models have yielded conflicting results. The effect of cobalamin inactivation by nitrous oxide on S-adenosylmethionine and S-adenosylhomocysteine has been examined in fruit bats and pigs. In pigs with symptoms of subacute combined degeneration, neural S-adenosylmethionine concentrations were normal but S-adenosylhomocysteine concentrations were greatly raised, causing a significant depression of the methylation ratio (Weir et al 1988). By contrast, in symptomatic fruit bats regional brain concentrations of S-adenosyl-methionine and S-adenosylhomocysteine were found to be normal — as was the methylation ratio (van der Westhuyzen and Metz 1983; Vieira-Makings et al 1990). These conflicting findings may be due to interspecies variation, because the experimental designs were similar, as were the methods used.

In mice treated with cycloleucine there is a rise in brain methionine and a fall in brain S-adenosylmethionine (Lee et al 1992), and a reduction in the methylation of arginine$_{107}$-myelin basic protein (Crang and Jacobson 1980). These results were confirmed in chicks (Small and Carnegie 1982). In rats treated with cycloleucine there was also a reduction in the methylation of arginine$_{107}$-myelin basic protein, but, using the same methodology, no reduction was seen in rats with cobalamin inactivated by nitrous oxide (Deacon et al 1986).

In summary, there is considerable interspecies variation in the findings. Pigs with inactivated cobalamin and mice, rats and chicks with inhibition of methionine adenosyltransferase all show defects in methylation; rats and fruit bats with inactivated cobalamin do not. Rats do not develop subacute combined degeneration after cobalamin inactivation, yet fruit bats do. Such variation indicated the need for further studies in humans, the most fruitful of which have been the study of inborn errors of metabolism.

Human studies — the role of inborn errors of metabolism

Inborn errors of cobalamin metabolism: The intracellular metabolism of cobalamin is complex and is summarized in Figure 2. Cobalamin bound to transcobalamin-2 is taken up by cells through the mechanism of receptor-mediated endocytosis. The complex dissociates in lysosomes and free cobalamin is released into the cytoplasm; here it is the fully oxidized cob(III)alamin. Cob(III)alamin is then reduced to cob(II)alamin in the cytoplasm and is either transported into the mitochondria or bound onto methionine synthase in the cytoplasm. In the mitochondria, cob(II)alamin is reduced to cob(I)alamin and then acquires an adenosyl group to give the cofactor adenosylcobalamin. In the cytoplasm, cob(II)alamin bound onto methionine synthase is reduced to cob(I)alamin and acquires a methyl group to form the holoenzyme complex methylcobalamin–methionine synthase. Inborn errors of metabolism can affect each step of the intracellular metabolism of cobalamin, and these are shown in Figure 2.

In mammals, cobalamin has been unequivocally identified as the cofactor for two enzymes only, methylmalonyl-CoA mutase, which requires adenosylcobalamin, and methionine synthase, which requires methylcobalamin. Methylmalonyl-CoA mutase is important in the catabolism of odd-chain fatty acids and the accumulation of these

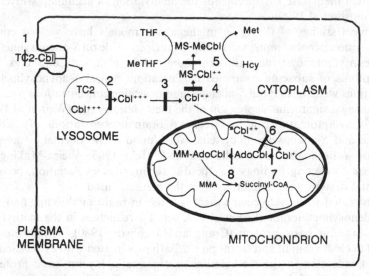

Figure 2 Intracellular cobalamin metabolism: Cbl is cobalamin; TC2 is transcobalamin-2; Ado is deoxyadenosyl; MM is methylmalonyl-CoA mutase; MMA is methylmalonyl-CoA; MS is methionine synthase; Me is methyl; THF is tetrahydrofolate; Hcy is homocysteine; Met is methionine. The numbers refer to the presumed site of the block in known inborn errors of metabolism in this pathway: 1 is transcobalamin-2 deficiency; 2 is Cbl F, defective lysosomal efflux of free cobalamin; 3 is Cbl C and Cbl D, defects in cob(III)alamin reductase; 4 is Cbl E, a defect in methionine synthase; 5 is Cbl G, a defect in the reduction and methylation of the methionine synthase holoenzyme; 6 is Cbl A, a defect in cob(II)alamin reductase; 7 is Cbl B, a defect in adenosyltransferase; 8 is mut⁰ or mut⁻, a defect in apomethylmalonyl-CoA mutase

in myelin had been suggested as a mechanism causing the demyelination (Vivaqua et al 1966; Frenkel 1971). However, patients with inborn errors of adenosylcobalamin alone (cbl A and cbl B) do not develop demyelination, neither do patients with absent or reduced methylmalonyl-CoA mutase apoenzyme (mut[0] or mut[-]) (Rosenberg and Fenton 1989). This provides compelling evidence that deficiency of adenosylcobalamin does not produce demyelination.

In contrast, demyelination has been demonstrated in children with inborn errors of methylcobalamin metabolism alone or combined methyl- and adenosylcobalamin defects. Typical demyelination has been demonstrated at post-mortem in one child with a combined defect (Dayan and Ramsey 1974), and has been demonstrated in life in children with defects of methylcobalamin alone by the characteristic clinical findings and by magnetic resonance imaging (Carmel et al 1988; Surtees et al 1991). This suggests that it is deficiency of methylcobalamin that is important in the development of the demyelination, in agreement with experimental findings (Metz 1992).

Inborn errors of the methyl-transfer pathway: If deficiency of methylcobalamin is critical to the development of demyelination, then the crucial substrate deficiency might be identified by examining serial inborn errors of the methyl-transfer pathway. A patient with 5,10-methylenetetrahydrofolate reductase deficiency causing central nervous system methyltetrahydrofolate deficiency was found to have very low cerebrospinal fluid concentrations of S-adenosylmethionine at diagnosis; these rose towards normal with treatment (Hyland et al 1988). This provided the first evidence for a link between S-adenosylmethionine and demyelination in humans. A larger study has examined the relationship between cerebrospinal fluid concentrations of 5-methyltetrahydrofolate, methionine and S-adenosylmethionine and demyelination in patients with inborn errors of methylfolate and methylcobalamin metabolism and methionine adenosyltransferase deficiency (Surtees et al 1991); the biochemical findings have also been contrasted with β-cystathionine deficiency (where demyelination does not occur) (Surtees 1992). Demyelination was associated *only* with deficiency of S-adenosylmethionine. Treatment caused remyelination demonstrable on neuroimaging and a rise of S-adenylmethionine concentrations into the normal range. It appears that S-adenosylmethionine deficiency is critical to the development of the demyelination in these inborn errors.

CONCLUSIONS

Current evidence suggests that, in humans, deficiency of S-adenosylmethionine is critical to the development of demyelination in cobalamin deficiency. However, the experimental and human data are conflicting. Crucial to the interpretation of the human data is the demonstration that cerebrospinal fluid concentrations of the methyl-transfer metabolites reflect those in the brain and the spinal cord. This problem has been addressed to some extent by examining the behaviour of cerebrospinal fluid concentrations of these metabolites in humans under conditions where, experimentally, direct measurements have been made in the neural tissue of rats (Surtees and Hyland 1990a,b). Under these circumstances, the behaviour of the cerebrospinal fluid metabolites has exactly mirrored that in the experimental animals. Yet, in mature pigs, direct comparison of cerebrospinal fluid and neural concentrations of S-adenosylmethionine has failed to show a good correlation (Weir et al 1992).

While the study of inborn errors of metabolism is important in focusing upon methionine synthase as a crucial enzyme involved in the maintenance of myelin, much further work needs to be done, especially to support the idea that deficiency of S-adenosylmethionine is critical in the development of demyelination in cobalamin deficiency. In particular it is crucial that a deficiency of this compound is demonstrated in brain and spinal cord tissue in humans with cobalamin deficiency.

REFERENCES

Adams RD, Kubik CS (1944) Subacute degeneration of the brain in pernicious anemia. *N Engl J Med* **231**: 1–9.

Amur SG, Shanker G, Cochran JM, Ved HS, Pieringer RA (1986) Correlation between inhibition of myelin basic protein (arginine) methyltransferase by sinefungin and lack of compact myelin formation in cultures of cerebral cells from embryonic mice. *J Neurosci Res* **16**: 367–376.

Auletta A, Gery A, Parmar A, Davis J, Mishra L, Mead J (1974) The effect of folate and folate analogues upon dihydrofolate reductase and DNA synthesis in kidneys of normal mice. *Life Sci* **14**: 1541–1553.

Beckman DR, Hoganson G, Berlow S, Gilbert EF (1987) Pathological findings in 5,10-methylenetetrahydrofolate deficiency. *Birth Defects* **23**: 47–64.

Billings F (1902) The changes in the spinal cord and medulla in pernicious anemia. *Boston Med Surg J* **147**: 225–233, 257–263.

Blackley R (1969) *The Biochemistry of Folic Acid and Related Pteridines.* Amsterdam: North-Holland.

Cantoni GL (1975) Biological methylation: selected aspects. *Annu Rev Biochem* **44**: 435–451.

Carmel R, Watkins D, Goodman SL, Rosenblatt DS (1988) Hereditary defect of cobalamin metabolism (Cbl G mutation) presenting as a neurological disorder in adulthood. *N Engl J Med* **318**: 1738–1741.

Clayton PT, Smith I, Harding B, Hyland K, Leonard JV, Leeming RJ (1986) Subacute combined degeneration of the cord, dementia and Parkinsonism due to an inborn error of folate metabolism. *J Neurol Neurosurg Psychiatr* **49**: 920–927.

Crang AJ, Jacobson W (1980) The methylation in vitro of myelin basic protein by arginine methylase from mouse spinal cord. *Biochem Soc Trans* **8**: 611–612.

Dayan AD, Ramsey RB (1974) An inborn error of vitamin B12 metabolism associated with cellular deficiency of coenzyme forms of the vitamin. Pathological and neurochemical findings in one case. *J Neurol Sci* **23**: 117–128.

Deacon R, Purkiss P, Green R, Perry J, Chanarin I (1986) Vitamin B12 neuropathy is not due to failure to methylate myelin basic protein. *J Neurol Sci* **72**: 113–117.

Frenkel EP (1971) Studies on the mechanism of the neural lesion of pernicious anemia. *J Clin Invest* **30**: 33a–34a.

Ghosh SK, Syed SK, Jung S, Paik WK, Kim S (1990) Substrate specificity for myelin basic protein-specific protein methylase I. *Biochim Biophys Acta* **1039**: 142–148.

Hoffman DR, Marion DW, Cornatzer WE, Duerre JA (1980) S-adenosylmethionine and S-adenosylhomocysteine metabolism in isolated rat liver. Effects of L-methionine, L-homocysteine and adenosine. *J Biol Chem* **255**: 10822–10827.

Hyland K, Smith I, Bottiglieri T et al (1988) Demyelination and decreased S-adenosylmethionine in 5,10-methylenetetrahydrofolate reductase deficiency. *Neurology* **38**: 459–462.

Israëls MCG, Wilkinson JF (1949) Risk of neurological complications in pernicious anaemia treated with folic acid. *Br Med J* **2**: 1072–1075.

Layzer RB, Fishman RA, Schaffer JA (1978) Neuropathy following abuse of nitrous oxide. *Neurology* 28: 504–506.

Lee C-C, Surtees R, Duchen L (1992) Distal motor axonopathy and central nervous system vacuolation caused by cycloleucine, an inhibitor of methionine adenosyltransferase. *Brain* 115: 935–955.

Lever EG, Elwes RDC, Williams A, Reynolds EH (1986) Subacute combined degeneration of the cord due to folate deficiency: response to methyl-folate treatment. *J Neurol Neurosurg Psychiatr* 49: 1203–1207.

Levitt M, Nixon P, Pincus J, Bertino J (1971) Transport characteristics of folate in cerebrospinal fluid; a study utilising doubly labelled 5-methyltetrahydrofolate and 5-formyltetrahydrofolate. *J Clin Invest* 50: 1301–1308.

Metz J (1992) Cobalamin deficiency and the pathogenesis of nervous system disease. *Annu Rev Nutr* 12: 59–79.

Lichtheim (1887) Zur kenntnis der perniciösen Anämie. *Verhandl d Cong f Innere Med* 6: 84–96.

Pallis CA, Lewis PD (1974) *The Neurology of Gastrointestinal Disease*. London: Saunders, 30.

Ramsey RB, Fischer VW (1978) Effect of 1-aminocyclopentane-1-carboxylic acid (cycloleucine) on developing rat central nervous system phospholipids. *J Neurochem* 30: 447–457.

Rosenberg LE, Fenton WA (1989) Disorders of propionate and methylmalonate metabolism. In Scriver CR, Beaudet AL, Sly WS, Valle D, eds. *The Metabolic Basis of Inherited Disease*, 6th edn. New York: McGraw-Hill, 821–844.

Russell JSR, Batten FE, Collier J (1900) Subacute combined degeneration of the spinal cord. *Brain* 23: 39–110.

Schatz RA, Wilens TE, Sellinger OZ (1981) Decreased transmethylation of biogenic amines after in vivo elevation of brain *S*-adenosyl-L-homocysteine. *J Neurochem* 36: 1739–1748.

Scott JM, Dinn J, Wilson P, Weir DG (1981) Pathogenesis of subacute combined degeneration: a result of methyl group deficiency. *Lancet* 2: 334–337.

Small DH, Carnegie PR (1982) In vivo methylation of an arginine in chicken myelin basic protein. *J Neurochem* 38: 184–190.

Small DH, Carnegie PR, Anderson RMcD (1981) Cycloleucine-induced vacuolation of myelin is associated with inhibition of protein methylation. *Neurosci Lett* 21: 287–292.

Smith I, Hyland K, Kendall B, Leeming R (1985) Clinical role of pteridine therapy in tetrahydrobiopterin deficiency. *J Inher Metab Dis* 8 (Suppl 1): 39–45.

Surtees R (1992) *S*-Adenosylmethionine deficiency and demyelination. A study of the metabolites of the methyl-transfer pathway in cerebrospinal fluid. PhD thesis, University of London.

Surtees R, Hyland K (1990a) Cerebrospinal fluid concentrations of *S*-adenosylmethione, methionine and 5-methyltetrahydrofolate in a reference population: cerebrospinal fluid *S*-adenosylmethionine declines with age in humans. *Biochem Med Metab Biol* 44: 192–199.

Surtees R, Hyland K (1990b) L-3,4-Dihydroxyphenylalanine (levodopa) lowers central nervous system *S*-adenosylmethionine concentrations in humans. *J Neurol Neurosurg Psychiatr* 53: 569–572.

Surtees R, Leonard J, Austin S (1991) Association of demyelination with deficiency of cerebrospinal fluid *S*-adenosylmethionine in inborn errors of methyl-transfer pathway. *Lancet* 338: 1550–1554.

Tabor CW, Tabor H (1976) 1,4-Diaminobutane (putrescine), spermidine and spermine. *Annu Rev Biochem* 45: 285–306.

Ungley C (1949) Subacute combined degeneration of the cord. I. Response to liver extracts. II. Trials with vitamin B12. *Brain* 72: 382–427.

van der Westhuyzen J, Fernandes-Costa F, Metz J (1982) Cobalamin inactivation by nitrous oxide produces severe neurological impairment in fruit bats: protection by methionine and aggravation by folates. *Life Sci* 31: 2001–2010.

van der Westhuyzen J, Metz J (1983) Tissue *S*-adenosyl-methionine levels in fruit bats (*Rousettus aegyptiacus*) with nitrous oxide-induced neuropathy. *Br J Nutr* 50: 325–330.

Vieira-Makings E, Metz J, van der Westhuyzen J, Bottiglieri T, Chanarin I (1990) Cobalamin neuropathy. Is S-adenosylhomocysteine toxicity a factor? *Biochem J* **266**: 707–711.

Vivaqua RJ, Myerson RM, Prescott DJ, Rabinowitz JL (1966) Abnormal propionic–methyl malonic–succinic acid metabolism in vitamin B12 deficiency and its possible relationship to the neurologic syndrome of pernicious anaemia. *Am J Med Sci* **251**: 507–515.

Weir DG, Keating S, Molloy A et al (1988) Methylation deficiency causes vitamin B12-associated neuropathy in the pig. *J Neurochem* **51**: 1949–1952.

Weir DG, Molloy AM, Keating JN et al (1992) Correlation of the ratio of S-adenosyl-L-methionine to S-adenosyl-L-homocysteine in the brain and cerebrospinal fluid of the pig: implications for the determination of this methylation ratio in human brain. *Clin Sci* **82**: 93–97.

Woltman HW (1918) Brain changes associated with pernicious anaemia. *Arch Int Med* **21**: 791–843.

Zappia V, Salvatore F, Porcelli M, Cacciapuoti G (1979) Novel aspects in the biochemistry of adenosylmethionine and related sulphur compounds. In Zappia V, Usdin E, Salvatore F, eds. *Biochemical and Pharmacological Roles of Adenosylmethionine and the Central Nervous System*. Oxford: Pergamon Press, 1–16.

J. Inher. Metab. Dis. 16 (1993) 771–786
© SSIEM and Kluwer Academic Publishers. Printed in the Netherlands

Inborn Errors and Demyelination: MRI and the Diagnosis of White Matter Disease

B. E. KENDALL

*Hospital for Sick Children, Great Ormond Street and National Hospital for
Neurology and Neurosurgery, Queen Square, London WC1N 3BG, UK*

Summary: The progress and extent of myelination can be assessed using magnetic resonance imaging (MRI). Myelination is delayed or diminished in several inherited metabolic abnormalities presenting in early life. Only minimal myelination of the CNS occurs in Pelizaeus–Merzbacher disease. Dysmyelination tends to produce fairly symmetrical lesions affecting white matter. In many mitochondrial enzyme and some lysosomal defects, the grey matter is also involved. The appearances and in particular the distribution on MRI and/or CT are characteristic in some conditions and the diagnosis is limited in others. Demyelination due to inflammatory disorders typically causes multifocal white matter lesions, recurrent in multiple sclerosis, monophasic in acute disseminated encephalomyelitis, extending in progressive multifocal leukoencephalopathy and classically involving the pons or corpus callosum in myelinolysis. Hypoxic ischaemic lesions may be metabolically induced and simulate primary demyelinating disorders. Mitochondrial enzyme defects in particular may present with stroke-like appearances. In many of these conditions, diagnosis is biochemical, but imaging has a significant role in suggesting the diagnosis, and documenting progression, response to therapy or complications.

MAGNETIC RESONANCE IMAGING AND MYELINATION

Myelin contains layered cholesterol and glycolipid combined with large protein molecules (Brown 1984). The T_1 of cholesterol is short and the T_1 of water is reduced by protein, so that as myelination progresses the intensity of white matter on T_1-weighted images increases. Myelin lipids are hydrophobic, so that loss of water accompanies myelination and this causes reduction in proton density and in signal on T_2-weighted images. The reduction in T_1 precedes the reduction in T_2 relaxation and proton density. A multislice optimal bandwidth double-echo STIR/TR/TI/TE-3000/148/23/85 utilizing second-order gradient motion rephasing on the longer echoes and an asymmetrical field of view presents mainly T_1-based signal at 23 ms and high contrast dependent on both T_1 and T_2 at 85 ms, and it can be imaged so as to show all the changes accompanying myelination as decreasing signal: the sequence can be optimized to produce excellent contrast between normal and pathological structures.

Hence the progress of myelination of the central nervous system can be documented by using appropriate MRI sequences. The MRI appearances of normal myelination reflect histological events with a delay of under 2 weeks.

Normal myelination: Myelination commences during the 5th month of gestation and continues throughout life. Myelin of Schwann cell origin is the first to be formed: it is practically confined to the cranial and peripheral nerves but does extend for a short distance into the central nervous system. Within the central nervous system, myelination is related to oligodendrocyte activity: in the brain myelination progresses in general inferosuperiorly, posteroanteriorly and centrifugally.

However, the process of myelination is not always homogeneous and it may be slightly asymmetrical: relative delay in completion of myelination is usual in the deep parietal white matter.

Myelination is evident: (1) At birth in the spinal cord, medulla oblongata, cerebellar peduncles, dorsal midbrain, posterior limbs of the internal capsules and ventrolateral thalami. (2) By 3 months in cerebellum, pons, anterior limbs of the internal capsules. (3) By 4 months in splenium of the corpus callosum and occipital white matter. (4) By 6 months in genu of corpus callosum, central centrum semiovale. (5) By 12 months in frontal white matter. (6) By 18 months throughout the cerebral hemispheres.

Delayed myelination: Delayed myelination is demonstrated by MRI in about 10% of retarded and brain-damaged infants. It is also a feature of malnutrition (Barkovich and Jackson 1989) and myelination is resumed and will proceed to completion when nutrition is corrected. Some inherited abnormalities of metabolism presenting in the first 12–18 months of life have diminished myelination: these include pyruvate dehydrogenase deficiency; propionic acidaemia, in which damage which may progress to necrosis is often evident in basal ganglia (Figure 1); and untreated phenylketonuria, in which neuronal migration defects also may be evident. Cerebral white matter hypoplasia is an idiopathic condition associated with delayed myelination in addition to generalized diminution in the volume of white matter.

Myelination on MRI is absent or minimal in the central nervous system in the type of X-linked Pelizaeus–Merzbacher disease (Figure 2) that presents in infancy with abnormal eye movements, head tremor, cerebellar ataxia and retardation. Dysmyelination and demyelination are well shown by MRI within myelinated neural substance, but cannot be recognized before myelination has occurred.

Dysmyelination: The imaging abnormalities induced by dysmyelination and inherited disorders of metabolism tend to be bilateral and more or less symmetrical, though there are notable exceptions. One classification of dysmyelination is based on current knowledge of the part played by organelles in the pathogenesis of hereditary disorders. Arrangement in this way has the advantage of reflecting some gross histological features common to some of the conditions in the groups. In some conditions CT and MRI have characteristic appearances and the differential diagnosis is limited in others (Kendall 1992). Atrophy supervenes in many conditions as the disease processes advance.

Lysosomal storage disorders: In metachromatic leukodystrophy, gliosis accompanying diffuse demyelination of the deep hemispheric white matter is followed by diffuse atrophy. These changes are reflected in neuroimaging (Figure 3). Similar appearances

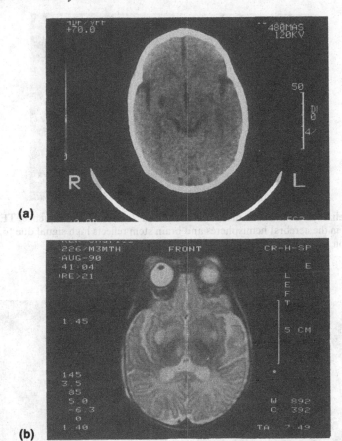

Figure 1 Boy aged 3.5 months; pyruvate dehydrogenase deficiency. (a) There is symmetrical low density in putamina of the lentiform nuclei. (b) MRI STIR sequence; TE 85 ms. There are symmetrical high signal foci in the lentiform nuclei and in both thalami. Myelination is slightly delayed: some myelin is present in the posterior but none has yet appeared in the anterior limb of the internal capsule

may occur in postinfective or postimmunization encephalitis, diffuse small-vessel disease, and as one manifestation of mitochondrial cytopathy in which the intermediate cerebral white matter tends to be more severely affected (Figure 4).

In globoid leukodystrophy, cerebroside accumulation in histiocytes may relate to high CT density in the thalami noted typically in cases presenting in infancy: this may be confused on CT with the after-effects of profound perinatal asphyxia, but the conditions are easily distinguished on history. CT high density in basal ganglia and thalami with low T_1 and high T_2 signal on MRI may also occur in the GM_1 and GM_2 gangliosidoses (Figure 5) due to accumulation of gangliosides and secondary demyelination. Low density and altered signal in the cerebral white matter become pronounced as the diseases advance. In the early stages, the affected parts of the brain are swollen and there may be secondary macrocrania. The swelling is later

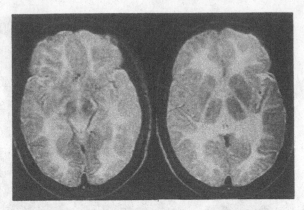

Figure 2 Pelizaeus–Merzbachr disease in a child of 2 years. STIR sequence; TE 85 ms. The white matter in the cerebral hemispheres and brain stem reflects high signal due to the absence of myelination

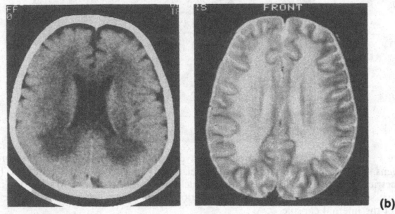

(a)   **(b)**

Figure 3 Metachromatic leukodystrophy. (a) CT section through corona radiata. There is symmetrical low density throughout the deep cerebral white matter. There is also mild cerebral atrophy. An incidental cavum septum pellucidi is shown. (b) MRI from a different patient. STIR sequence; TE 85 ms. There is symmetrical high signal throughout the cerebral white matter with the exception of arcuate fibres so that the periphery is spared

replaced by atrophy (Fukumizu et al 1992).

In globoid leukodystrophy of later clinical onset the predominant feature is gliosis in hemispheric white matter, induced by toxic effects of psychosine. This causes much less specific symmetrical CT low density and T_1 low, T_2 high signal, mainly in the parietal regions: similar appearances may occur, for example, in the gangliosidoses, peroxisomal disorders and phenylketonuria.

In mucopolysaccharidoses types I, II, III and VI, storage products accumulate in distended perivascular spaces and the enlarged spaces, which on CT and MRI appear

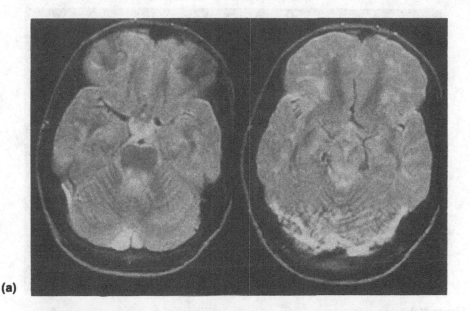

(a)

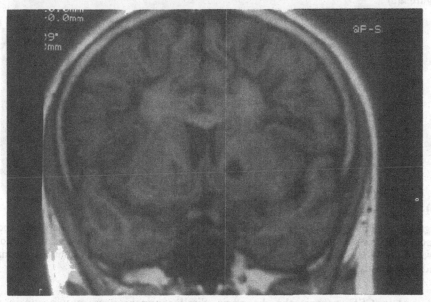

(b)

Figure 4 Mitochondrial cytopathy in a 13-year-old male. (a and b) Axial STIR sequence; TE 85 ms. Normal low signal is returned from the myelinated deep hemispheric white matter, internal capsules and pons. Abnormal high signal is returned from the peripheral hemispheric white matter and midbrain. There is also high signal from the lentiform nuclei and substantia nigra. (c) Coronal T_1-weighted sequence. Normal signal return from the corpus callosum, deep hemispheric white matter and internal capsules. The abnormal peripheral white matter and the lentiform nuclei return abnormally low signal

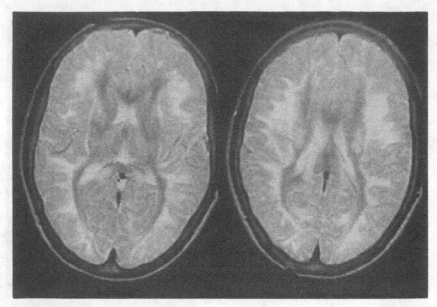

Figure 4(c)

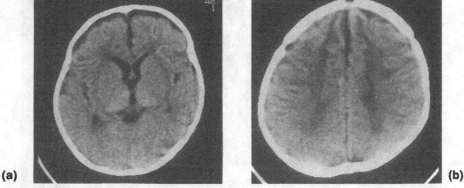

(a)   (b)

Figure 5 Infantile GM₂ gangliosidosis in a boy aged 1 year. Retarded, cherry-red spot at macula, hepatosplenomegaly. Cranial CT scan. (a) Section through thalami. (b) Section at level of roof of lateral ventricles. There is symmetrical low density in the frontal white matter and high density in both thalami, consistent with the diagnosis

to contain CSF, suggest the diagnosis. Similar appearances may occur in very different clinical contexts in cryptococcal meningitis and rarely in Lowe syndrome. Meningeal infiltration in the mucopolysaccharidoses may cause hydrocephalus and subdural effusions, and ligamentous infiltration may result in cord compression. The latter is, however, a more prominent feature of mucopolysaccharidosis IV, in which there is vertebral dysplasia and commonly instability with antlanto-axial subluxation and os odontoideum formation.

J. Inher. Metab. Dis. 16 (1993)

Peroxisomal disorders: In the adrenoleukodystrophy–adrenomyeloneuropathy group (Figure 6), there is characteristic involvement of the peritrigonal white matter, splenium of the corpus callosum, posterior limb of the internal capsule and parts of the crus cerebri: the geniculate bodies, brachia and cerebellar white matter also may be abnormal. Reactive changes to demyelination and necrosis result in breakdown of the blood–brain barrier and inflammatory cellular infiltration where the disease process is active. This is reflected on imaging in a characteristic band of enhancing tissue (Figure 7) close to the outer margin of the disease process, which is usually

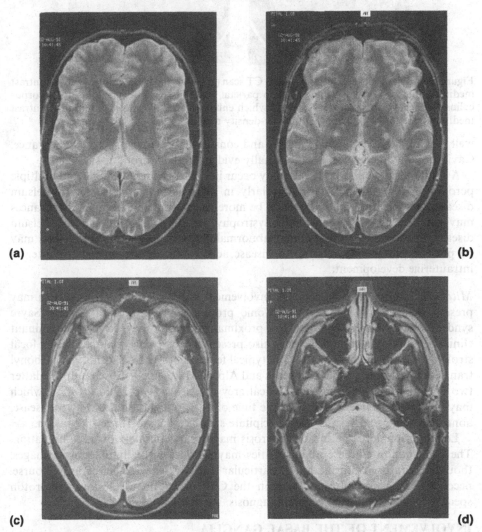

(a)

(b)

(c)

(d)

Figure 6 Adrenomyeloneuropathy. STIR sequence; TE 85 ms. (a–d) High signal involves the peritrigonal white matter, the splenium of the corpus callosum, the posterior limbs of the internal capsules, the crus cerebri, and the cerebellar white matter

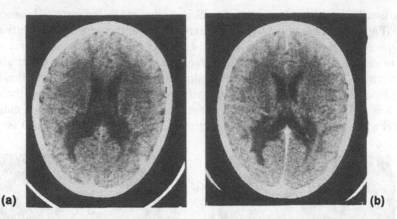

Figure 7 Adrenoleukodystrophy. Cranial CT scan (a) before and (b) after intravenous contrast medium. There is low density in the deep parietal white matter and splenium of the corpus callosum. There is a higher-density band which enhances after the administration of contrast medium, close to the periphery of the low-density region

well seen in the cerebral white matter and constitutes an almost specific appearance. Cavitation or calcification is occasionally evident in the necrotic tissue.

A similar distribution of disease may occur in the less common single and multiple peroxisomal enzyme defects, particularly in infantile Refsum disease. In Refsum disease itself the abnormality tends to be more widespread and imaging appearances may simulate metachromatic leukodystrophy. In Zellweger and infantile Refsum disease, in addition to white matter abnormalities, neuronal migration defects may be present, presumably relating to disease activity during the second trimester of intrauterine development.

Mitochondrial enzyme dysfunction: Involvement of the central nervous system may present as a slowly progressive chronic process with features of Kearns–Sayre syndrome or of MERRF in which a proximal myopathy may be the predominant clinical effect. Some conditions may also present as acute encephalopathies or focal stroke-like syndromes. The latter are typical features of MELAS, ornithine carbonyl transferase deficiency, Menkes disease and Alpers–Huttenlocher disease. In the latter two conditions, degeneration of cortical grey matter leads to rapid atrophy, which may be the most evident feature at the time of first presentation. In Menkes disease, abnormal meningeal vessels may precipitate subdural haemorrhage or effusion.

Leukoencephalopathy and cell necrosis may occur separately or in combination. The combination of these abnormalities may produce virtually diagnostic images, though laboratory confirmation of particular causative enzyme defects is, of course, necessary. Detection of lactic acid in the CSF or within the lesions on proton spectroscopy supports the general diagnosis.

INVOLVEMENT OF THE BASAL GANGLIA
The basal ganglia tend to be diffusely involved in mitochondrial enzyme dysfunction and in acute episodes there may be considerable swelling, which may resolve or

progress to necrosis with cavitation (Figure 4) or calcification. Similar basal ganglion changes occur in glutaric aciduria type I, but in this condition there is also generally widening of the sylvian fissures and subarachnoid spaces over the convexity of the frontal lobes. In some cases, this appears to be due to atrophy; in others it is associated with expansion of the vault and suggests arachnoid adhesion with cyst formation. Atrophy of the basal ganglia with enlargement of the frontal horns is a frequent sequel in glutaric aciduria type I.

The putamina are particularly vulnerable to anoxia and to defects in respiratory chain enzymes. The subthalamic nuclei have tended to show signal changes in cytochrome oxidase deficiency and the globus pallidus in methylmalonic and propionic acidemia.

Neuronal loss with spongiform degeneration, demyelination, necrosis and cystic change in the basal ganglia occur in Wilson disease and result in T_1 hypointense, T_2 hyperintense lesions on MRI and low density on CT. In addition, T_2 low signal (Figure 8) may be evident, consistent with deposition of copper or its replacement by ferric iron, both of which cause loss of phase coherence with T_2 shortening.

Calcification in the basal ganglia and cerebral subcortical white matter together with dysmyelination of cerebral white matter predominantly in the posterior parts of the hemispheres should suggest malignant hyperphenylalaninaemia. Congenital AIDS may give rise to similar appearances.

BRAIN STEM ABNORMALITY

Involvement of the brain stem on MRI is a prominent feature in Leigh syndrome, particularly when older children and young adults are affected, maple syrup urine

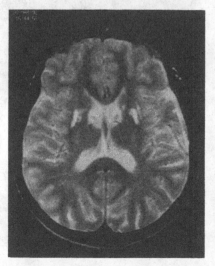 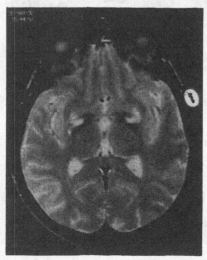

(a) (b)

Figure 8 Wilson disease. (a and b) T_2-weighted spin echo sequence, contiguous sections. There are foci of high signal in the lentiform nuclei and heads of the caudate nuclei. There is also a little abnormal low signal in the lentiform nuclei due to the dephasing caused by deposition of a paramagnetic substance which could be either ferric iron or copper

disease, partial albinism with immune insufficiency (PAID) (Brismar et al 1992) and Wilson disease. In all these conditions there may be swelling that may resolve and be replaced by atrophy.

LESIONS SIMULATING INFARCTS

In MELAS, large single or multiple lesions involving both white and grey matter in continuity, and thus resembling infarcts, may affect any part of the brain but are more frequent in the posterior parts of the cerebral hemispheres. If other features of mitochondrial cytopathy, such as basal ganglion calcification or brain stem or cerebellar atrophy are present, the images should suggest the diagnosis. The other conditions causing stroke-like events are diagnosed biochemically: in ortnithine transcarbamylase deficiency high levels of glutamine may be shown within the lesions on proton spectroscopy.

VACUOLATING MYELOPATHY

Canavan disease is the commonest of a group of conditions characterized by splitting of myelin and accumulation of fluid within or between the layers of the sheaths, producing a vacuolating myelopathy. Extensive white matter abnormality with early involvement of the cortical U-fibres suggests the diagnosis (Figure 9). The fluid extends into and distends gyral cores, causing macrocephaly and macrocrania. Deep

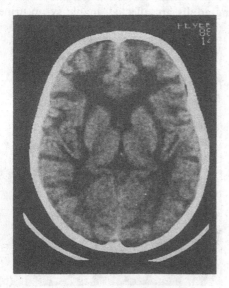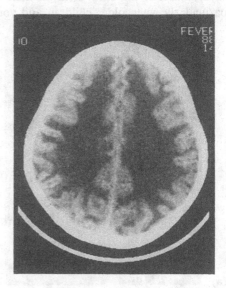

Figure 9 Canavan disease. Child aged 2 years with macrocephaly and increasing regression. Axial CT scan. There is symmetrical low density throughout the white matter of both cerebral hemispheres, extending into the cores of the gyri and the external capsules. The internal capsules are spared.

extension of the abnormality on imaging is predictive of a relatively rapid downhill clinical course. In most cases, a typical proton spectrum is produced due to accumulation of N-acetylaspartic acid, but similar histological appearances occur in the absence of aspartoacylase deficiency. In such idiopathic cases, the clinical course tends to be more benign and may arrest: in addition to the diffuse white matter abnormality, imaging may show focal cystic changes, particularly in the high medial posterior frontal and parietal white matter. Congenital muscular dystrophy of the Fukuyama type shows delayed myelination and often migrational abnormalities. It may also show cerebral changes similar to Canavan disease with macrocephaly: less extensive changes, sometimes multifocal, occur in some organic acidopathies.

INVOLVEMENT OF THE CEREBELLUM

In the presence of widespread cerebral involvement, the cerebellum is frequently abnormal in many inherited metabolic abnormalities as, for example, in Canavan disease, the peroxisomal enzyme deficiencies, Wilson disease and PAID. Isolated symmetrical cerebellar white matter abnormality may occur in cerebrotendinous xanthomatosis (Figure 10). It also occurs in Langerhans' cell histiocytosis, usually in association with diabetes insipidus due to pituitary stalk infiltration (Figure 11) and in carbohydrate-deficiency glycoprotein syndrome.

Disorders of DNA repair, characteristically manifest in Cockayne disease, are reflected in extensive leukoencephalopathy, associated with cellular damage, calcification and gross atrophy with cerebellar and brain stem involvement (Demaerel et al 1992).

ALEXANDER DISEASE

In some leukodystrophies in which a metabolic defect has not been discovered the appearances on imaging combined with clinical abnormalities can suggest the

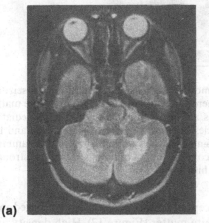
(a)
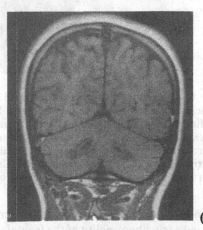
(b)

Figure 10 Cerebrotendinous xanthomatosis. Sections through cerebral hemispheres: (a) axial T_2-weighted, (b) coronal T_1-weighted. There is abnormal signal return from the white matter around and within the hilum of the dentate nuclei. There was no further abnormality

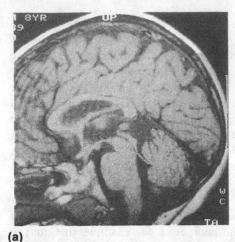 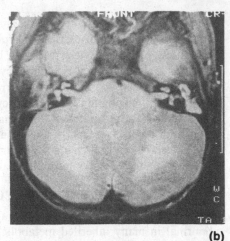

(a)   (b)

Figure 11 Langerhans histocytosis in an 8-year-old boy with previous bone lesion; developed diabetes insipidus and mild ataxia. (a) Midsagittal T_1-weighted section. There is thickening of the pituitary stalk and of the floor of the hypothalamus in the region of the tuber cinereum. The normal high signal from the posterior pituitary is absent owing to lack of posterior pituitary hormones. (b) T_2-weighted axial section. There is abnormal high signal returned from the white matter of the cerebellar hemispheres

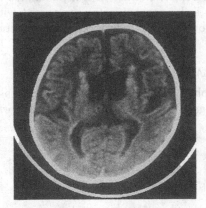

Figure 12 Alexander disease in a boy aged 13 months. Increasing retardation with seizures and increasing macrocrania. Cranial CT scan. There is low density in the frontal white matter extending through the external capsules. There is abnormal high density in the immediately subependymal regions around the lateral ventricles, the heads of the caudate nuclei and the proximal parts of the forceps minor. The lateral ventricles, the sylvian fissures and the anterior convexity subarachnoid spaces are enlarged. The combination of radiological findings strongly supports the diagnosis, which was confirmed by histology

diagnosis. Thus, in Alexander disease, macrocephaly may be associated with predominant and early abnormality of the frontal white matter (Figure 12). High density and abnormal enhancement is commonly shown on CT in the forceps major and minor and around the lateral ventricles early in the evolution of the disease. Focal swelling and cavitation or calcification may uncommonly be shown, particularly in the frontal white matter (Clifton et al 1991). In the juvenile-onset subgroup, enlargement of the lower medulla with signal change in the posterior part has been observed.

FOLLOW-UP STUDIES

Re-imaging, by showing the progression of disease, may also aid in diagnosis and the recognition of complications in cases where clinical progression is unexpectedly rapid. Resolution with or without progression to atrophy, especially with clinical improvement, is uncommon in most leukodystrophies and may suggest an alternative diagnosis, such as a reversible metabolic deficiency or encephalitis.

DEMYELINATING DISEASES

The second major group of white matter disorders are *acquired*. These may be divided into: (1) a group characterized by peri-axonal demyelination with relative preservation of axons; and (2) a group in which demyelination may be a major feature, but axons as well as myelin are primarily involved.

The former group includes inflammatory white matter disorders, the predominant condition being multiple sclerosis, which may be subdivided into several clinical types with differences on imaging. It is a multifocal disease in which hypodense areas may be shown on CT, but MRI is much more frequently abnormal and typically reveals multiple foci of T_2 high signal in white matter, the periventricular regions being most commonly affected. Enhancement, which may encompass a lesion or ring the periphery, occurs in late lesions, and usually persists for up to 4 weeks.

Acute disseminated perivenous encephalomyelitis (Figure 13) (ADEM) and acute

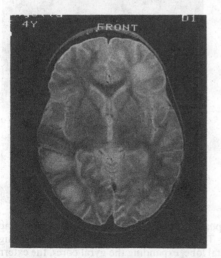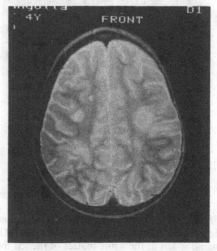

Figure 13 Acute disseminated encephalomyelitis in a child aged 4 years. Ten days after respiratory tract infection developed paraparesis with CSF lymphocytosis. Partial spontaneous recovery. No further episodes over 18 months observation. Cranial MRI: T_2-weighted sections. There are multiple high signal lesions without mass effect involving the cerebral white matter

haemorrhagic leukoencephalitis produce similar appearances (Kesselring et al 1990) but are monophasic disorders. They may not be distinguished at the onset of the disease, but new lesions appearing after 6 months exclude ADEM and suggest multiple sclerosis. MRI has made substantial contributions not only to diagnosis but also to our understanding of the natural history of these disorders. Progressive multifocal leukoencephalitis, due to infection of oligodendroglia by papova viruses in immune-compromised patients, may cause typical appearances on computed imaging (Figure 14). There are focal lesions, often multiple, mainly or exclusively involving white matter, extending into gyral cores and along tracts. There is only minor mass effect and generally, but not invariably, enhancement is absent.

Some acquired metabolic deficiencies and toxic conditions give rise to characteristic, though not pathognomonic distributions of white matter change. Included among these are myelinolysis typically affecting the central pons (Figure 15), but often more widespread or occasionally exclusively extrapontine. Pontine myelinolysis is associated with electrolyte imbalance: it occurs typically in severe alcoholism and other forms of malnutrition, but it may complicate hepatic insufficiency, including that of Wilson disease. The corpus callosum may be affected in myelinolysis, but is classically the site of Marchiafava–Bignami disease that complicates chronic alcoholism (Figure 16). In the acute phase the corpus callosum is swollen with T_2 high signal and CT low density. Within a few weeks, the swelling abates and the signal intensity becomes heterogeneous and within months this is replaced by callosal atrophy and cavitation (Chang et al 1992).

Exogenous toxins, including certain organic solvents, may primarily affect myelin;

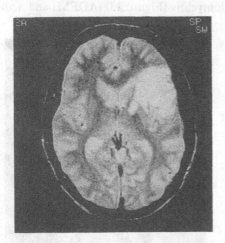 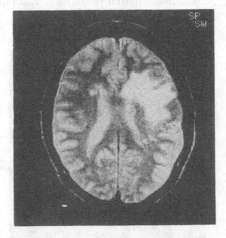

Figure 14 Progressive multifocal leukoencephalitis in a drug abuser with AIDS. Rapid deterioration with hemiparesis. Cranial MRI; T_2-weighted sections. There is a large high-signal lesion in the white matter of the left frontal lobe, expanding the gyral cores, the external capsule and the anterior limb and genu of the internal capsule. The lesion also extends into the immediately adjacent parts of the grey matter. For its size, the lesion has little mass effect. There are further small foci of high signal in the left parietal and right frontal white matter. There was no abnormal enhancement.

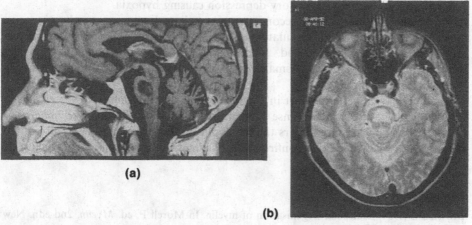

(a)

(b)

Figure 15 Central pontine myelinolysis in an alcoholic. Presented with malnutrition and electrolyte imbalance which was corrected. This was followed by tetraparesis and a locked-in syndrome. MRI: (a) sagittal midline T_1-weighted section, (b) axial T_2-weighted section. There is extensive abnormal signal from the central pons (T_1 low, T_2 high) sparing the periphery. The appearances are classical for central pontine myelinolysis. There is cerebellar atrophy consistent with the effects of high alcohol intake

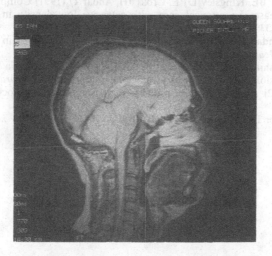

Figure 16 Marchiafava–Bignami disease in a 64-year-old with history of alcoholism. Developed dementia and delusions. Sagittal T_2-weighted section. Shows increased signal confined to the corpus callosum. The appearances support a clinical diagnosis. (No histological confirmation)

others including cocaine and heroin act indirectly, the former through cerebrovascular spasm and the latter by respiratory depression causing hypoxia.

Important conditions in the second group as well as degenerative or hypertensive small-vessel disease include circulatory disorders of hypoxic ischaemic aetiology *in utero* or in the perinatal period. Inflammatory, postradiation and metabolically induced (as in Fabry disease) small-vessel narrowing may also simulate primary demyelinating disorders.

Imaging plays a valuable part in the recognition of white matter diseases and in following the course and response to therapy. In some conditions, the imaging is characteristic and in many others the metabolic and/or inflammatory nature of the disease can be suggested, for confirmation by clinical correlation, biochemical tests or histology.

REFERENCES

Brown PE (1984) Molecular organisation of myelin. In Morell P, ed. *Myelin*, 2nd edn. New York: Plenum, 97–116.

Barkovich AJ, Jackson DE Jr (1989) MRI assessment of normal and abnormal brain myelination. *MRI Divisions* 19–25.

Brismar J, Harfi HA (1992) Partial albinism with immunodeficiency: a rare syndrome with prominent white matter changes. *AJNR* **13**: 387–393.

Chang KH, Cha SH, Han MH, Park SH, Nah Dh, Hong JH (1992) Marchiafava–Bignami disease: serial changes in corpus callosum on MRI. *Neuroradiology* **34**: 480–482.

Clifton A, Kendall BE, Kingsley DPE, Cross JH, Andar U (1991) Computed tomography in Alexander's disease: An atypical case with extensive low density in both frontal lobes. *Neuroradiology* **33**: 438–440.

Demaerel Ph, Kendall BE, Kingsley D (1992) Cranial CT and MRI in diseases with DNA repair defects. *Neuroradiology* **34**: 117–121.

Fukumizu M, Yoshikawa H, Takashima S, Sakuragawa N, Kurokawa T (1992) Tay–Sachs disease: progression of changes on neuroimaging in four cases. *Neuroradiology* **34**: 483–496.

Kendall BE (1992) Disorders of lysosomes, peroxisomes, and mitochondria. *AJNR* **13**: 621–653.

Kesselring J, Miller DH, Robb SA et al (1990) *Brain* **113**: 291–302.

J. Inher. Metab. Dis. 16 (1993) 787–790

Genetic Analysis of Batten Disease

R. M. GARDINER

Department of Paediatrics, University College London Medical School, Rayne Institute, University Street, London, WC1E 6JJ, UK

Summary: Batten disease, or neuronal ceroid-lipofuscinosis (CLN) comprises a group of inherited neurodegenerative disorders characterized by the accumulation of autofluorescent lipopigment in neurones. The three main childhood varieties — infantile (CLN1), late-infantile (CLN2) and juvenile (CLN3) — manifest autosomal recessive inheritance. The basic biochemical defect remains unknown. The strategy of positional cloning is being pursued to elucidate the molecular basis of Batten disease. The infantile disease locus (CLN1) has been mapped by linkage analysis to human chromosome 1p32, and the juvenile disease locus (CLN3) to human chromosome 16p12. In each case marker loci in strong linkage disequilibrium with the disease loci have been identified. Locus heterogeneity between classical late-infantile CLN (CLN2) and both CLN1 and CLN3 has been demonstrated. Work is in progress to clone CLN1 and CLN3 and to map CLN2. Identification of linked markers has provided a new • approach to prenatal diagnosis. The methodology exists for positional cloning of these genes and elucidation of the molecular genetic basis of the ceroid lipofuscinoses.

The strategy of positional cloning has been successfully applied to an increasing number of Mendelian diseases in man. Families segregating for the disease are studied with multiple polymorphic markers until identification of linkage allows localization of the disease gene. The region containing the gene can then usually be narrowed down to a genetic interval of about one centimorgan (cM). If no cytogenetic clues are available it is then necessary to analyse the candidate region — a million base pairs or more of DNA — for all transcripts, and investigate these for an alteration confined to affected individuals.

The first phase of this strategy has been accomplished for the infantile (CLN1) and juvenile (CLN3) ceroid-lipofuscinosis disease loci. Available evidence indicates that the classical late-infantile disease (CLN2) arises from mutations at a third locus. Recent work on the genetic analysis of these three diseases is considered in turn.

INFANTILE NEURONAL CEROID-LIPOFUSCINOSIS — CLN1

This is one of the so-called Finnish diseases. The age of onset is 8–18 months. The storage material appears as granular osmiophilic deposits. Inheritance is autosomal recessive and the disease is enriched in the Finnish population with an incidence of 1 : 20 000.

A random search of the genome was initiated. Initially, only 6 CLN1 families with 2 affected children or fetuses were available. After analysing 42 DNA and protein polymorphisms, a positive LOD-score was found using a VNTR marker, D15S57, at 1p (Jarvela et al 1991). The localization was subsequently refined to 1p32 in the vicinity of the L-myc gene (Jarvela 1991).

Recent multipoint analyses identify the most likely location of CLN1 as 0.2 cM proximal to a new tetranucleotide repeat marker (GAAA)$_n$ lying 16 kb upstream from the L-myc gene (Hellsten et al, in press).

Allelic association: Strong allelic association has been observed between CLN1 and three markers on 1p: L-myc, (GAA)$_n$ and D1S62 (Jarvela 1991; Hellsten et al, in press). Finnish disease chromosomes possess an allele of this marker which is present on less than 3% of chromosomes from the family resource of the Centre for the Study of Human Polymorphism (CEPH), or of the non-disease chromosomes of the CLN1 family parents. Eighty-seven percent of Finnish patients are homozygous for this rare allele, suggesting that only one major mutation causes infantile CLN in Finland.

Prenatal and carrier diagnosis: Before the assignment of CLN1 to chromosome 1, prenatal diagnosis was carried out by electron microscopy of a chorionic villus biopsy (Rapola et al 1990). Prenatal diagnosis using the linked DNA markers has been carried out for 18 fetuses at risk. In all cases there has been concordance between DNA investigation and EM results (Jarvela et al 1991). The existence of a highly informative marker showing strong allelic association has significantly improved risk calculations for both prenatal and carrier diagnosis.

JUVENILE NEURONAL CEROID LIPOFUSCINOSIS — CLN3

Juvenile-onset CLN usually presents with visual failure at age 4–8 years. Macular and retinal degeneration occurs. The characteristic storage material appears as 'fingerprint profiles' on electron microscopy. Lymphocytes show vacuolation.

Genetic linkage studies using 'classical' protein polymorphisms identified linkage to the haptoglobin locus, allowing assignment of CLN3 to human chromosome 16 (Eiberg et al 1989). Subsequent studies using additional families and markers allowed refinement of the localization to the region 16p12 (Gardiner et al 1990; Callen et al 1991).

In the most recent analysis, 16 polymorphic loci were typed in a total of 72 families. Pairwise LOD-scores at these loci are shown in Table 1. Analysis of multiple informative recombinant meioses locate CLN3 in the interval between D16S297 and D16S57.

Allelic association: Within the above interval, alleles at three loci — D16S288, D15S298 and D16S299 — show strong association with the disease locus (Mitchison et al, in press). It is likely that CLN3 is within one megabase of these loci. Prenatal

Table 1 Two-point LOD scores between CLN3 and chromosome 16 marker loci in 66 CLN3 pedigrees

	Maximum LOD score (z)	Recombination fraction (θ), sex-averaged
CL N3 versus		
D16S159	4.75	0.10
D16S294	3.36	0.05
D16S167	7.03	0.06
D16S67	10.27	0.02
D16S295	9.44	0.02
D16S296	14.03	0.02
D16S297	16.91	0.01
D16S148	6.30	0.00
D16S298	23.18	0.00
D16S299	24.00	0.00
D16S57	4.87	0.02
D16S285	16.98	0.02
D16S150	1.84	0.10
D16S151	0.59	0.19

Table 2 Pairwise LOD scores between CLN3 and CLN2 and the microsatellite marker D16S298

	Recombination fraction						
	0.00	0.01	0.05	0.10	0.20	0.30	0.40
CLN3-D16S298	23.18	22.51	20.36	16.81	10.23	4.85	1.27
CLN2-D16S298	−99.00	−17.59	−7.18	−3.50	−0.97	−0.23	−0.03

diagnosis of juvenile CLN using linked DNA markers has been reported (Uvebrant et al, in press).

CLASSICAL LATE-INFANTILE NEURONAL CEROID LIPOFUSCINOSIS — CLN2

Onset is between ages of 2 and 4 years. Ultrastructural examination typically demonstrates curvilinear bodies. Patients have been described with features intermediate between the classical late-infantile and juvenile varieties.

The locus for classical late-infantile CLN (CLN2) has been excluded from the regions to which CLN1 and CLN3 map. Multipoint linkage analysis of CLN2 and eight marker loci on chromosome 1p allowed exclusion of a region of about 40 cM as the site of CLN2.

Pairwise linkage analysis of CLN2 and the marker locus D16S298 provided exclusion data for a region of at least 20 cM (Table 2). In addition, there was no evidence of linkage disequilibrium between alleles at this locus and CLN2.

REFERENCES

Callen DF, Baker E, Lane S et al (1991) Regional mapping of the Batten disease locus (CLN3) to human chromosome 16p12. *Am J Hum Genet* **49**: 1372–1377.

Eiberg H, Gardiner RM, Mohr J (1989) Batten disease (Spielmeyer–Sjogren disease) and haptoglobins (HP): indication of linkage and assignment to chromosome 16. *Clin Genet* **36**: 217–218.

Gardiner RM, Sandford A, Deadman M et al (1990) Batten disease (Spielmeyer–Vogt disease, juvenile onset neuronal ceroid lipofuscinosis) gene (CLN3) maps to human chromosome 16. *Genomics* **8**: 387–390.

Hellsten E, Vesa J, Jarvela I, Makela TP, Santavuori P, Peltonen L (1993) Refined assignment of the infantile neuronal ceroid lipofuscinosis locus at 1p32 and the current status of prenatal and carrier diagnostics. *J Inher Metab Dis* **16**: 335–338.

Jarvela I (1991) Infantile neuronal ceroid lipofuscinosis (CLN1): linkage disequilibrium in the Finnish population and evidence that variant late-infantile form (variant CLN2) represents a non-allelic locus. *Genomics* **10**: 333–337.

Jarvela I, Rapola J, Peltonen L et al (1991a) *Prenat Diagn* **11**: 323–328.

Jarvela I, Schleutker J, Haataja L et al (1991b) Infantile form of neuronal ceroid lipofuscinosis (CLN1) maps to the short arm of chromosome 1. *Genomics* **9**: 170–173.

Mitchison HM, Williams RE, McKay TR et al (1993) Refined genetic mapping of juvenile onset neuronal ceroid lipofuscinosis on chromosome 16. *J Inher Metab Dis* **16**: 339–341.

Rapola J, Salonen R, Ammala P, Santavuori P (1990) Prenatal diagnosis of the infantile type of neuronal ceroid lipofuscinosis by electron microscopic investigation of human chorionic villi. *Prenat Diagn* **10**: 553–559.

Uvebrant P, Bjorck E, Conradi N, Hokegard K-H, Martinsson T, Wahlstrom J (1993) Successful DNA-based prenatal exclusion of juvenile neuronal ceroid lipofuscinosis. *Prenat Diag* in press.

J. Inher. Metab. Dis. 16 (1993) 791–799
© SSIEM and Kluwer Academic Publishers. Printed in the Netherlands

Recent Developments in Menkes Disease*

H. KODAMA
Department of Pediatrics, Teikyo University School of Medicine, 11-1, Kaga 2, Itabashi-ku, Tokyo 173, Japan

Summary: Recent studies on Menkes disease are reviewed, focusing especially on copper transport in the cells. A large amount of copper accumulated in the organelle-free cytoplasm, whereas mitochondria were in a state of copper deficiency, indicating that Menkes mutation probably affects copper transport from the cytosol to the organelles in the cells. Microscopic observation of the brain of the macular mouse showed that copper accumulates in the blood vessels. Observation of the brain tissue of the macular mouse after intraventricular administration of copper revealed that copper accumulates in the glia as well as the blood vessels. Copper accumulation was also observed in cultured astrocytes, a type of glial cell, indicating that the affected astrocytes accumulate blood-borne copper and release little of it in the patients with Menkes disease. Thus the effective treatment of Menkes disease could possibly be to release trapped copper from the blood vessels and glia into the neurons.

Menkes disease is an X-linked recessive disorder described first by Menkes and colleagues (1962) in which patients show characteristic symptoms such as lack of keratinization and pigmentation of hair and neuronal degeneration. Danks and colleagues (1972) demonstrated that Menkes disease is an inborn error of copper metabolism in which intestinal copper absorption is disturbed. This fact, together with the discovery of model animals such as the brindled mouse (Hunt 1974) and the macular mouse (Nishimura 1975) led to many studies on copper metabolism in this disease. These studies revealed that most ingested copper in the diet accumulates in the intestine; copper accumulation is also observed in other organs and cells such as the kidney and cultured fibroblasts. The excessive copper is bound to metallothionein in the cells, the level of which is higher than in the control cells. Defective intestinal absorption of copper results in copper deficiency in the blood, liver and brain, which is the main clinical feature of this disease (Danks 1989). The gene and the primary metabolic defect of this disease, however, have not yet been

---

*Editors' note. Since the submission of this paper a candidate gene for Menkes disease has been identified by three independent groups (Chelly et al 1993; Mercer et al 1993; Vulpe et al 1993). The gene product established from the nucleotide sequence is highly homologous with the P-type cation-transporting ATPases and in particular with a bacterial copper-transporting ATPase. Some patients with Menkes disease have been shown to have partial gene deletions and in others the expression of the gene transcript has been altered or reduced.

elucidated. An effective treatment for this disease has also not yet been established. In this paper, recent studies on copper metabolism in this disease are reviewed, focusing especially on copper transport in the affected cells and the brain.

DIAGNOSIS AND GENETICS

The diagnosis of patients with Menkes disease can be made with great confidence by the clinical features and laboratory findings. In the case of inherited diseases such as Menkes disease, however, prenatal diagnosis and diagnosis of carriers are important. The prenatal diagnosis of Menkes disease is currently made by copper analysis of cultured amniotic cells or chorionic villus. However, it has been pointed out that maternal deciduum contamination of a chorionic villus sample possibly gives a false-positive diagnosis, because copper content in the maternal deciduum is high when the pregnant woman is a carrier (Tønnesen et al 1989). ^{64}Cu uptake analysis of amniotic cells can also be useful in the diagnosis of affected fetuses. Tønnesen and colleagues (1989) reported that a combination of ^{64}Cu uptake with chase experiments on the amniotic cells can give a much better distinction between affected and unaffected fetuses than ^{64}Cu uptake studies alone. However, there is so far no definitive method for prenatal diagnosis of Menkes disease.

The Menkes gene has recently been reported to be located distal to the X-inactivation centre and proximal to the phosphoglycerate kinase 1 locus in Xq13 (Tümer et al 1992; Tønnesen et al 1992). Although the exact localization of the Menkes gene has not yet been established, it is reported that if flanking DNA markers of the Menkes locus are used for heterozygote detection and prenatal diagnosis, the risk of misclassification will be less than 1% (Tønnesen et al 1992).

LOCALIZATION OF COPPER IN THE AFFECTED CELLS

In Menkes disease, copper accumulates in various tissues, especially in the intestine. In an attempt to find the localization of copper in the affected cells, we examined histochemically the site of excess copper accumulation in the intestine and kidney of macular mice (Kodama et al in preparation). Figure 1 shows the electron microscopic localization of copper in the intestinal epithelium of the macular mouse. An excess of copper is concentrated in the organelle-free cytoplasm of the epithelial cells. Copper accumulation is also observed in the vascular endothelium. These observations suggest that ingested copper is trapped and accumulated in the cytoplasm of epithelial cells and in the vascular endothelium. As shown in Figure 2, copper was detected more intensely in the organelle-free cytoplasm of the proximal tubular cells of the kidney. In this case also, copper was not detected in the nucleus, mitochondria and lysosome (Figure 2). These results show that copper accumulation occurs in the organelle-free cytoplasm of the affected cells. Analysis of the copper level of mitochondria obtained from cultured fibroblasts of patients with Menkes disease revealed that mitochondria were in a state of copper deficiency although the cells accumulated excess amounts of copper (Table 1) (Kodama et al 1989). Therefore, Menkes mutation probably affects copper transport from the cytosol to the organelles in the cells; that is, copper transport in the cytosol is disturbed in this disease.

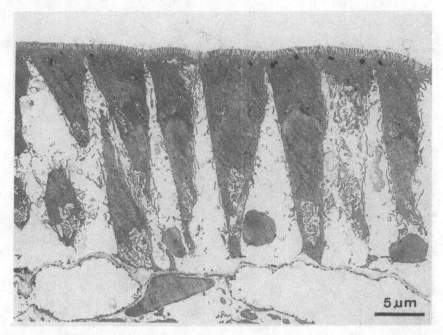

Figure 1 Electron microscopic observation of the intestine of a macular mouse stained for copper

PATHOGENESIS

In affected cells the excess cytosolic copper is known to be bound to metallothionein. Thus the level of metallothionein is also higher in the affected cells. However, the regulatory mechanism of metallothionein synthesis in response to intracellular copper concentration in the affected cells is reported to be normal (Packman et al 1987). Therefore, it is probable that a high level of metallothionein in the affected cells is induced by a large amount of cytosol copper. Increased amounts of metallothionein in the affected cells seem to be a secondary consequence of the mutation (Mercer et al 1991).

The activity of superoxide dismutase 1, another cytosolic cuproprotein, in the affected cells is reported to be in the normal range (Packman et al 1984), indicating that cytosolic copper in the cells is available for normal binding to cuproproteins. On the other hand, Palida and Ettinger (1991) reported that the level of a 48-kDa cytosolic copper-binding protein that is suggested to be involved in copper transport in the cytosol decreases or disappears in both hepatic and renal cytosols from the brindled mouse, and suggested that the primary defect of Menkes disease might be due to the decreased level of this protein. If the gene of this protein is located on Xq13, a decrease or disappearance of this protein could be involved in the basic defect in Menkes disease. Further studies on this hypothesis are needed.

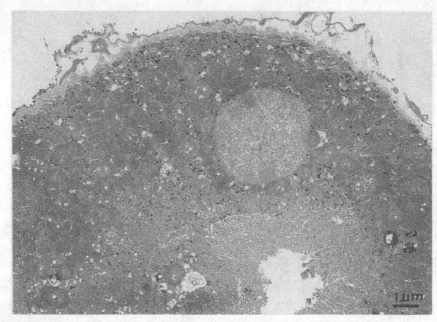

Figure 2 Electron-microscopic observation of the proximal tubular cells in the kidney from a macular mouse stained for copper

Table 1   Copper concentration in cultured fibroblasts from patients with Menkes disease

	Copper (ng/mg protein)[a]	
	Intracellular	*Mitochondrial*
Menkes disease	365 ± 70 ($n = 6$)	44 ± 20
Control	107 ± 21 ($n = 8$)	78 ± 11

From Kodama et al (1989)
[a]Mean $\pm$ SD

COPPER TRANSPORT IN THE BRAIN

The most characteristic symptom of Menkes disease is neuronal degeneration, which is considered to be caused by copper deficiency in the brain. It is well known that administration of copper after birth does not improve the patients' neuronal degeneration (Johnsen et al 1991). In order to resolve this problem, the distribution of copper in the brain of macular mice was examined (Kobayashi et al unpublished observations). Figure 3 shows the microscopic examination of cerebellar tissue of a 14-day-old macular mouse and of a normal littermate stained for copper. The choroid plexus and ependyma are intensely stained in both the macular mouse and the control mouse. However, the glia of the control is intensely stained, whereas that of the macular mouse is faintly stained. The neurons of the control are strained faintly, but

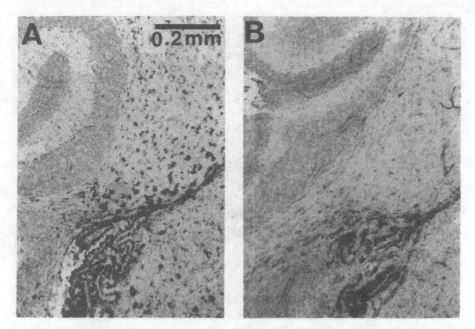

Figure 3 Distribution of copper in the cerebellum of a macular mouse: (Λ) normal littermate; (B) macular mouse. These sections are lightly counterstained with toluidine blue

those of the macular mouse are not stained. In contrast, the blood vessels of the macular mouse are stained intensely, whereas no staining is observed in those of the control, showing that the blood vessels of the brain of the macular mouse abnormally accumulate copper.

Brain tissue of the macular mice was observed after administration of copper directly into the ventricle from the brain surface. Glia was stained intensely in this case, showing that glia of the macular mouse accumulates copper (Figure 4). As described above, glia of untreated macular mouse is not stained for copper. Seemingly, copper is not delivered to the glia because of abnormal copper accumulation in the blood vessels in the brain of the macular mouse. When the astrocytes, a type of glial cell, were cultured, those of macular mice accumulated larger amounts of copper than control cells (Kodama et al 1991). This indicates that the genetic defect of the macular mouse is expressed in the astrocytes. Probably the affected astrocytes accumulate blood-borne copper and release little of it in patients with Menkes disease. Figure 5 shows a hypothesis to explain the mechanism of copper transport derived from the observations described above. In normal animals, blood-borne copper is transported to neurons via the vascular endothelium and the astrocytes. In patients with Menkes disease, however, blood-borne copper is trapped in the blood vessels and the astrocytes, and thus is not transported to neurons. If this is the case, it accords well with the fact that intravenously administered copper does not improve the neuronal degeneration.

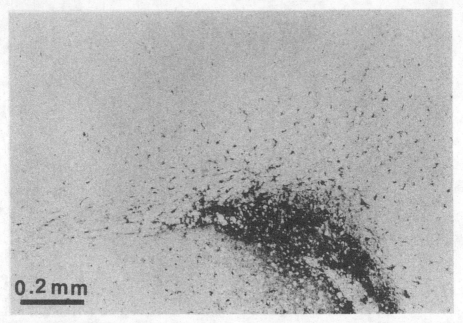

Figure 4 Brain section stained for copper after administration of copper directly into the ventricle of a macular mouse. Glial cells of the corpus callosum near the lateral ventricle are intensely stained

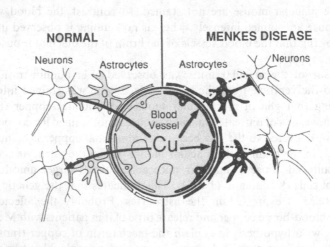

Figure 5 Hypothesis for copper transport in the brain

TREATMENT

As described above, parenteral administration of copper into patients with Menkes disease after birth produces no improvement of the neuronal degeneration of patients. In contrast, macular mice are known to survive and show almost normal growth when they are administered copper about postnatal day 7. Gradual improvement of the neuropathological findings and of cuproenzyme activity is also observed (Yamano et al 1988; Meguro et al 1991). The 7th postnatal day of the mouse is known to correspond with the mid-third trimester of humans. Thus the brain of the mouse on that day is still immature. The blood–brain barrier of the mouse on that day is possibly immature enough to allow administered copper to enter the brain. This means that the parenterally administered copper may be available to the neurons in patients with Menkes disease and the animal model only when their astrocytes are immature. These considerations suggest that an effective treatment to improve the neuronal degeneration of the patients is copper administration to the fetus. However, it is difficult practically to treat the affected fetuses with copper.

Postnatal treatments currently applied are parenteral copper administration in combination with either D-penicillamine (Nadal and Baerlocher 1988), chelators (Tanaka et al 1990) or vitamin E (Tada et al 1988), treatment with vitamin C (de Groot et al 1989), and treatment with parenteral copper-histidinate (Sherwood et al 1989; Kollros et al 1991). However, it is not clear whether these treatments are effective. Blood-borne copper has been found to be trapped by the blood vessels and the glial cells in the brain. Effective treatment for releasing the trapped copper from the blood vessels and glial cells to the neurons should be elucidated in the future.

CLINICALLY ATYPICAL MENKES PATIENTS AND OTHER PATIENTS WITH COPPER DEFICIENCY

Patients with Menkes disease show characteristic symptoms of copper deficiency. However, patients who show clinically different symptoms from classical Menkes patients are reported. For example, Haas and colleagues (1981) described two male patients who had no hypothermia and survived longer than usual. The ^{64}Cu uptake and retention values on fibroblast cultures from these so-called milder forms are reported to be indistinguishable from those of classical Menkes patients (Tønnesen et al 1991). Longer survival has also been reported in classical Menkes patients (Gerdes et al 1989). These observations suggest that Menkes disease exhibits wide variety in the clinical picture. In addition, occipital horn syndrome and other copper deficiency diseases that are different from Menkes disease are reported (Fujii et al 1991; Iwakawa et al 1992). The elucidation of these diseases from various aspects, especially that of gene analysis, will provide knowledge on the primary defect of copper metabolism in these diseases that could provide a clue to solving the complex mechanism of copper metabolism in humans.

ACKNOWLEDGEMENTS

I express my sincere thanks to Dr S. Kobayashi, Dr I. Takahashi, Dr M. Nishimura, Ms Y. Meguro and Ms A. Tsunakawa for helpful advice and technical assistance.

This study was partly supported by grants from the Ministry of Health and Welfare of Japan and grants from the Ministry of Education, Science and Culture of Japan.

REFERENCES

Chelly J, Tümer Z, Tønnesen T et al (1993) Isolation of a candidate gene for Menkes disease that encodes a potential heavy metal binding protein. *Nature Genetics* **3**: 14–19.

Danks DM (1989) Disorders of copper transport. In Scriver CR, Beaudet AL, Sly WS, Valle D, eds. *The Metabolic Basis of Inherited Disease*, 6th edn. New York: McGraw-Hill, 1411–1431.

Danks DM, Campbell PE, Steevens BJ, Mayne V, Cartwright E (1972) Menke's kinky hair syndrome: An inherited defect in copper absorption with widespread effects. *Pediatrics* **50**: 188–120.

Fujii T, Okuno T, Ito M et al (1991) Non-Menkes-type copper deficiency with regression, lactic acidosis, and granulocytopenia. *Neurology* **41**: 1263–1266.

Gerdes AM, Tønnesen T, Pergament E et al (1989) Variability in clinical expression of Menkes syndrome. *Eur J Pediatr* **148**: 132–135.

de Groot CJ, Wijburg FA, Barth PG (1989) Vitamin C treatment in Menkes' disease: Failure to affect biochemical and clinical parameters. *J Inher Metab Dis* **12** (Suppl 2): 389–392.

Haas RH, Robinson A, Evans K, Lascells PT, Dubowitz V (1981) An X-linked disease of the nervous system with disorders of copper metabolism and features differing from Menkes disease. *Neurology* **31**: 852–859.

Hunt DM (1974) Primary defect in copper transport underlines mottled mutants in the mouse. *Nature* **249**: 852–854.

Iwakawa Y, Shimohira M, Kohyama J, Kodama H (1992) Sibling cases of a degenerative neurological disease associated with hypocupremia and hypobetalipoproteinemia. *Eur J Pediatr*, in press.

Johnsen DE, Coleman L, Poe L (1991) MR of progressive neurodegenerative change in treated Menkes' kinky hair disease. *Neuroradiology* **33**: 181–182.

Kodama H, Okabe I, Yanagisawa M, Kodama Y (1989) Copper deficiency in the mitochondria of cultured skin fibroblasts from patients with Menkes syndrome. *J Inher Metab Dis* **12**: 386–389.

Kodama H, Meguro Y, Abe T et al (1991) Genetic expression of Menkes disease in cultured astrocytes of the macular mouse. *J Inher Metab Dis* **14**: 896–901.

Kollros PR, Dick RD, Brewer RJ (1991) Correction of cerebrospinal fluid copper in Menkes kinky hair disease. *Pediatr Neurol* **7**: 305–307.

Meguro Y, Kodama H, Abe T, Kobayashi S, Kodama Y, Nishimura M (1991) Changes of copper level and cytochrome *c* oxidase activity in the macular mouse with age. *Brain Dev* **13**: 184–186.

Menkes JH, Alter M, Steigleder GK, Weakley DR, Sung JH (1962) A sex-linked recessive disorder with retardation of growth, peculiar hair and focal cerebral and cerebellar degeneration. *Pediatrics* **29**: 764–779.

Mercer JFB, Stevenson T, Wake SA, Mitropoulos G, Camakaris J, Danks DM (1991) Developmental variation in copper, zinc and metallothionein mRNA in brindled mutant and nutritionally copper deficient mice. *Biochim Biophys Acta* **1097**: 205–211.

Mercer JFB, Livingston J, Hall B et al (1993) Isolation of a candidate gene for Menkes by positional cloning. *Nature Genetics* **3**: 20–25.

Nadal D, Baerlocher K (1988) Menkes' disease: long-term treatment with copper and D-penicillamine. *Eur J Pediatr* **147**: 621–625.

Nishimura M (1975) A new mutant mouse, macular (Ml). *Exp Animal* **24**: 185 (in Japanese).

Packman S, Chin P, O'Toole C (1984) Copper utilization in cultured skin fibroblasts of the mottled mouse: an animal model for Menkes kinky hair syndrome. *J Inher Metab Dis* **7**: 168–170.

Packman S, Palmiter RD, Karin M, O'Toole C (1987) Metallothionein messenger RNA regulation in the mottled mouse and Menkes kinky hair syndrome. *J Clin Invest* **79**: 1338–1342.

Palida FA, Ettinger MJ (1991) Identification of proteins involved in intracellular copper metabolism. *J Biol Chem* **266**: 4586–4592.

Sherwood G, Sarkar B, Kortsak AS (1989) Copper histidinate therapy in Menkes' disease: Prevention of progressive neurodegeneration. *J Inher Metab Dis* **12** (Suppl. 2): 393–396.

Tada H, Tanaka M, Inada E et al (1988) A patient with Menkes disease effecting treatment with tocopherol acetate. *No to Hattatsu* **20**: 514–516 (in Japanese).

Tanaka K, Kobayashi K, Fujita Y, Fukuhara C, Onosaka S, Min K (1990) Effects of chelators on copper therapy of macular mouse, a model animal of Menkes' kinky disease. *Res Commun Chem Pathol Pharmacol* **69**: 217–227.

Tümer Z, Tommerup N, Tønnesen T, Kreuder J, Craig IW, Horn N (1992) Mapping of the Menkes locus to Xq13.3 distal to the X-inactivation center by an intrachromosomal insertion of the segment Xq13.3-q21.1. *Hum Genet* **88**: 668–672.

Tønnesen T, Gerdes A-M, Damsgaard E et al (1989) First-trimester diagnosis of Menkes disease: Intermediate copper values in chorionic villi from three affected male fetuses. *Prenat Diagn* **9**: 159–165.

Tønnesen T, Garret C, Gerdes A-M (1991) High ^{64}Cu uptake and retention values in two clinically atypical Menkes patients. *J Med Genet* **28**: 615–618.

Tønnesen T, Petterson A, Kruse TA, Gerdes A-M, Horn N (1992) Multipoint linkage analysis in Menkes disease. *Am J Hum Genet* **50**: 1012–1017.

Vulpe C, Levinson B, Whitney S, Packman S, Gitschier J (1993) Isolation of a candidate gene for Menkes disease and evidence that it encodes a copper-transporting ATPase. *Nature Genetics* **3**: 7–13.

Yamano T, Shimada M, Onaga A et al (1988) Electron microscopic study on brain of macular mutant mouse after copper therapy. *Acta Neuropathol* **76**: 574–580.

J. Inher. Metab. Dis. 16 (1993) 800–811
© SSIEM and Kluwer Academic Publishers. Printed in the Netherlands

Proton Magnetic Resonance Spectroscopy Studies in Lactic Acidosis and Mitochondrial Disorders

J. H. CROSS, D. G. GADIAN*, A. CONNELLY and J. V. LEONARD
Institute of Child Health and The Hospital for Sick Children, Great Ormond Street, London WC1N 3JH, UK

Summary: Congenital lactic acidoses form a large group of disorders that are commonly associated with profound neurological dysfunction. Difficulties are frequently encountered in establishing a diagnosis, and the mechanisms underlying brain damage are poorly understood. We have performed proton magnetic resonance spectroscopy (^{1}H-MRS) on 24 patients under investigation for suspected metabolic disorder, and have compared the MRS observations of brain lactate with measurements of cerebrospinal fluid (CSF) lactate. We have shown good concordance between the two types of observation, confirming the value of the CSF measurements. Regional variations in brain lactate are detected in some cases, and these may help to elucidate the mechanisms underlying selective brain damage.

INTRODUCTION

The congenital lactic acidaemias are a heterogeneous group of disorders presenting in a variety of ways. Neurological dysfunction is common, but the exact mechanisms responsible for brain damage are not known. Several disorders of pyruvate metabolism and the respiratory chain have now been described (Robinson 1985; Trijbels et al 1988). However, diagnosis can be difficult as characteristic metabolites are not found. Blood lactate is not specific, and is not reliable as it can be affected by stress, muscle movement and even venepuncture (Kollee et al 1977). For this reason cerebrospinal fluid (CSF) lactate is widely used, but it is difficult to validate. In normal individuals, the literature suggests good correlation between plasma and CSF lactate, but patients have been reported in whom CSF lactate has been raised while plasma lactate is normal (Brown et al 1988). However, measurement of CSF lactate is itself only an indirect method of examining central nervous system involvement, and it is not known if it is an accurate predictor of abnormal lactate metabolism in the brain.

Proton magnetic resonance spectroscopy (^{1}H-MRS) can now be used for the non-invasive examination of brain metabolites *in vivo*. In the normal brain the dominant contributions to the spectra are from *N*-acetylaspartate, creatine + phosphocreatine, and choline-containing compounds; lactate is difficult to detect because of its relatively

---

*Correspondence

low concentration. However, in several disorders of the brain, lactate can be seen by MRS at elevated concentrations (Bruhn et al 1989; Detre et al 1991; Grodd et al 1991; Luyten et al 1990). We have carried out ^{1}H-MRS studies on a series of children under investigation for suspected metabolic disorders, and we have compared the MRS results with CSF lactate findings.

PATIENTS

We carried out ^{1}H-MRS studies on 24 children who were being investigated for a suspected metabolic disorder at the Hospital for Sick Children. Clinical details of the patients are summarized in Table 1. There were eight males and 16 females, age range 13 days to 11 years. Five children had pyruvate dehydrogenase deficiency (assayed by Dr G. K. Brown (Oxford) in skin fibroblasts in patients 1–4 and on muscle in patient 5). Patient 11 had biotinidase deficiency diagnosed on plasma assay (Dr K. Bartlett, Newcastle). Fructose 1,6-bisphosphatase deficiency was suspected from the clinical information in patient 14 and confirmed on enzyme assay in white cells (Mr A. Whitfield, London). Cytochrome oxidase deficiency was diagnosed on histochemistry of liver in patient 13 and of muscle in patient 15 (Professor B. Lake, London).

METHODS

1. Magnetic resonance spectroscopy

A 1.5 T Siemens whole body system was used, with a standard quadrature head coil. The children were examined under sedation according to the protocol of the Hospital for Sick Children, Great Ormond Street, as previously described (Shepherd et al 1990). Imaging was carried out using a double-echo short TI inversion recovery (DE-STIR) sequence (TR = 4 s, TI = 1.45 ms, TE = 23 ms and 85 ms) (Finn et al 1989). Using these images to select a volume of interest, spectra were obtained from $2 \times 2 \times 2$ cm cubes centred on the basal ganglia and occipital white matter, using a 90–180–180 spin echo technique (Ordidge et al 1985) with the three selective radiofrequency pulses applied in the presence of orthogonal gradients of 2 mT/m. Water suppression was achieved by pre-irradiation of the water resonance using a 90° Gaussian pulse with a 60 Hz bandwidth, followed by a spoiler gradient. TR was 1600 ms and TE was 135 ms. After global and local shimming, and optimization of the water suppression pulse, data were collected in 2–4 blocks of 128 scans. The time domain data were corrected for eddy-current induced phase modulation using non-water-suppressed data as a reference (Klose 1990). Exponential multiplication corresponding to 1-Hz line broadening was carried out prior to Fourier transformation, and a cubic spline baseline correction was performed.

Assessment of the presence or absence of a lactate signal above the background noise was made by two independent observers.

2. Measurement of cerebrospinal fluid lactate

Lumbar puncture was performed in all children under sedation. Cerebrospinal fluid was collected into fluoride heparin tubes (Teklab Ltd., Sacriston, Co. Durham) and

Table 1 Clinical details of patients

Patient Age	Diagnosis	Presentation	Neurology	CSF lactate (mmol/L)
1 (F) 10 m	Pyruvate dehydrogenase deficiency	Neonatal metabolic acidosis	Microcephaly, developmental delay	7.6
2 (M) 2 y 4 m	Pyruvate dehydrogenase deficiency	Neonatal metabolic acidosis	Microcephaly, developmental delay	7.2
3 (F) 2 y 10 m	Pyruvate dehydrogenase deficiency	Developmental regression age 7 m. Subsequent slow developmental progress with seizures	OFC 3rd centile. Pyramidal signs all limbs. Ataxia.	3.3 (4.0*)
4 (F) 2 m	Pyruvate dehydrogenase deficiency	Hypotonia, cyanosis day 6	Lethargy, unresponsive truncal hypotonia	8.5
5 (F) 7 m	Pyruvate dehydrogenase deficiency	FTT, Developmental delay	Poor head control, developmental delay	7.7
6 (M) 6 y	Leigh disease (no enzyme diagnosis)	Transient lt sided weakness 18 m. Intermittent	Unable to stand due to ataxia. Dystonic posturing lt leg. Transient gaze palsy.	3.6
7 (F) 4 y 7 m	Leigh disease (no enzyme diagnosis)	Development delay. Polyneuropathy at 18 m. Sibling affected.	Developmental delay. Ataxia. Hyperreflexia.	4.2
8 (F) 8 m	Congenital lactic acidosis (precise defect not defined)	Seizures age 8 wks, poor feeding	Microcephaly, developmental delay, spastic quadriplegia, squint.	8.4
9 (M) 4 m	Congenital lactic acidosis (precise defect not defined)	Respiratory arrest age 3.5 m. Metabolic acidosis	Abnormal eye movements. Seizures	4.8
10 (F) 9 m	Congenital lactic acidosis (precise defect not defined)	Tachypnoea 8 m. Developmental delay	Hypotonia, developmental delay	7.4
11 (M) 5 m	Biotinidase deficiency	Seizures age 2 m. FH	Hypotonia, unresponsive. Persistent grasp and moro.	5.7
12 (F) 6 wks	Congenital lactic acidosis (precise defect not defined)	Floppy and poor feeding 2 days. Metabolic acidosis	Abnormal eye movements. Tone normal	2.6

continued

FTT: Failure to thrive; OFC: Occipitofrontal head circumference; *initial presentation

Table 1 *continued*

Patient	Age	Diagnosis	Presentation	Neurology	CSF lactate (mmol/L)
13 (M)	2 m	Cytochrome oxidase deficiency	Vomiting day 3, lethargic, acidotic	Hypertonia, fisting	8.2
14 (M)	14 m	Fructose-1,6-bisphosphatase deficiency	Tachypnoea and unwell age 9 h. Hypoglycaemic	Normal	2.1
15 (F)	10 y 6 m	Cytochrome oxidase deficiency + cataracts	Cataracts age 6 y	Proximal myopathy, Normal IQ	2.0
16 (F)	3 m	Congenital lactic acidosis (precise defect not defined)	Lens opacities. Renal tubular acidosis day 2. Hyperammonaemia	No neurological deficit at presentation	5.1
17 (F)	13 d	Congenital lactic acidosis (precise defect not defined)	Acidosis in the neonatal period	Normal	6.9
18 (F)	5 m	Familial neurodegenerative disorder	Abnormal from birth. Seizures 2–3 wks	Microcephaly, hypertonia. Persistent moro reflex	1.3
19 (F)	11 y	Severe mental retardation	Seizures from 12m Hyperventilation	Microcephaly, profound developmental delay with regression	1.6
20 (F)	9 m	Williams syndrome	Feeding problems, dysmorphic. Aortic stenosis	Developmental delay, tremor	2.5
21 (M)	5 y 4 m	Juvenile Parkinson disease	Floppy 6m, tremor since 9m	Extrapyramidal signs	1.4
22 (F)	2 y 1 m	Myoclonic epilepsy	Balance problem age 20m, frequent falls.	Developmental delay.	0.8
23 (M)	11 m	Neurodegenerative disease	Seizures at 3 wks	Myoclonic jerks	1.1
24 (F)	1 y 9 m	Cerebellar ataxia	Developmental delay age 6 m	Hypotonia, not fixing. Developmental delay OFC 50th. Ataxia.	1.1

FTT: Failure to thrive; OFC: Occipitofrontal head circumference; *initial presentation

assayed within 30 min of collection or frozen immediately and stored at $-20°C$ until assayed which was within one working day of collection. Lactate was measured using a YSI model 23L lactate analyser (Yellow Springs Instrument Co., Yellow Springs, Ohio, USA) calibrated with an in-house standard of 10 mmol/L L-lactate. CSF lactates were measured on the same day as the MRS examination in nine of the children, within 24–48 hours in a further seven, within 48–72 hours in a further two, and within 3 days–1 week in five more children. In the final child (patient 3), CSF lactate was measured three months after the initial MRS examination, but on re-assessment the following year, CSF and MRS measurements were performed on the same day.

RESULTS

Figure 1 shows a normal ^1H spectrum from the basal ganglia of a 12-month-old girl, together with an image showing the region selected for spectroscopy. The dominant contributions to the spectrum are from N-acetylaspartate (NAA), creatine + phospho-creatine (Cr), and choline-containing compounds (Cho). The relative intensities of these signals vary with age (van der Knapp et al 1990; Connelly et al 1991), and there is also a regional dependence (Frahm et al 1989; Connelly et al 1991), variations that must be taken into account when making comparisons.

Figure 2 shows a spectrum from the basal ganglia of a child (patient 2) aged 2 years 4 months with a CSF lactate of 7.2 mmol/L. In addition to the NAA, Cr and Cho signals, an additional negative doublet is observed, centred at 1.32 ppm. This is characteristic of lactate under the particular MRS conditions chosen (i.e. at the echo time TE = 135 ms), and indicates that lactate is elevated in this region of the brain.

Spectra were obtained from the basal ganglia in all 24 children, and from occipital white matter in 16 of the children. In six of the patients, CSF lactate was in the normal range (1.6 mmol/L or less (Van Erven et al 1985)), and in none of these children was a convincing lactate signal detected by MRS, either in the basal ganglia or in occipital white matter.

Three children had CSF lactates in the range 1.7–2.5 mmol/L. Brain lactate was not convincingly detected by MRS in any of these children. One of these children (patient 14) had fructose- 1,6-bisphosphatase deficiency, with a significantly higher plasma lactate (3.8 mmol/L) compared to CSF lactate (2.1 mmol/L). He was neuro-logically intact.

Two children had CSF lactates in the range 2.6–3.9 mmol/L. Both had MRS examinations on two occasions. In one of these children, lactate was seen by MRS in both examinations, and the CSF lactate (2.6 mmol/L) was only measured on the second examination at 6 weeks of age. She was noted at this time to have generalised increased tone and abnormal eye movements. In the other child, CSF lactate was 3.6 mmol/L on both occasions. Lactate was only seen in his second MRS examination, at a time of neurological deterioration.

In the remaining 13 patients, the CSF lactate was increased to 4 mmol/L and above (range 4.0–8.5 mmol/L), and in all these children lactate was observed in the ^1H spectra of at least one of the regions examined. The basal ganglia region was examined in all of these children, and occipital white matter in nine. Eleven of the children showed a definite lactate signal in the basal ganglia, and eight of the nine

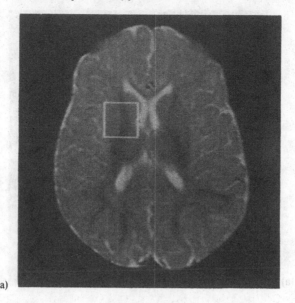

(a)

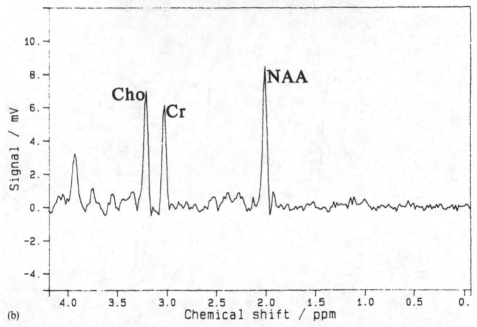

(b)

Figure 1 Magnetic resonance image (**a**) and a normal ^{1}H spectrum (**b**) obtained from a 12-month-old girl. The image shows slightly delayed myelination for age but is otherwise normal. The box indicates the position of the 8 ml cubic volume of interest from which the spectrum was obtained. The dominant contributions to the spectrum are from *N*-acetylaspartate (NAA), creatine + phosphocreatine (Cr) and choline-containing compounds (Cho).

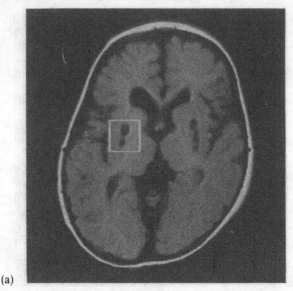

(a)

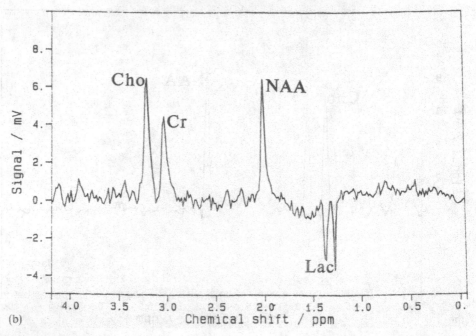

(b)

Figure 2 Magnetic resonance image (a) and ^1H spectrum (b) obtained from a child aged 2 years 4 months with a CSF lactate of 7.2 mmol/L (patient 2). The image shows generalized atrophy as well as some delay in myelination. The box indicates the position of the 8 ml cubic volume of interest from which the spectrum was obtained. The dominant contributions to the spectrum are from *N*-acetylaspartate (NAA), creatine + phosphocreatine (Cr), choline-containing compounds (Cho), and lactate (Lac).

examinations of occipital white matter also showed lactate. In the ninth examination of occipital white matter, interpretation was uncertain because of a contaminating signal from lipids which overlaps the signal from lactate. While regional variations in brain lactate were seen in some of the children, there was no obvious tendency for lactate to be consistently higher in one region than in the other.

The most striking example of regional differences is illustrated in Figure 3, which shows data from patient 3. This child was originally undergoing investigation for an undiagnosed leukodystrophy. The spectrum from the basal ganglia (Figure 3b) shows the characteristic NAA, Cr and Cho signals, with the possibility of an additional small signal from lactate. In comparison, the spectrum from abnormal occipital white matter (Figure 3d) shows gross abnormalities; in particular, there is overall loss of intensity of the NAA, Cr and Cho signals, and an extremely high signal from lactate. The observation of this high lactate signal provided the first biochemical guide to diagnosis, and in agreement with the MRS finding, the CSF lactate concentration was subsequently shown to be increased (4.0 mmol/L). This child was re-assessed one year later when she had made slow developmental progress but had developed seizures. MRS showed some increase in the NAA, Cr and Cho signals, and the lactate signals were similar to those observed in the first examination. CSF lactate on this occasion was 3.3 mmol/L, and enzyme analysis on cultured skin fibroblasts confirmed pyruvate dehydrogenase deficiency.

In principle, it is possible to determine absolute concentrations from the MRS signal intensities. However, this requires corrections for a variety of factors that could influence signal intensities, including T1 and T2 relaxation effects. In practice this is difficult to achieve within the time constraints imposed by clinical examinations, and we were not able to correct for all of these factors in the present studies. However, for patient 3 (Figure 3) the fact that the NAA, Cr and Cho signals were all significantly lower in the occipital white matter region than in the basal ganglia provides strong evidence that there is overall loss of cells in the white matter region. Furthermore, comparison with model solutions containing lactate indicated that the lactate concentration in this region must be at least 10 mmol/kg wet wt, and that it could possibly be considerably higher.

Further information is available from the *N*-acetylaspartate (NAA) signal. While the function of NAA remains uncertain, it is believed to be located primarily within neurons, and therefore an unusually low NAA in comparison with age- and region-matched normal data is commonly interpreted in terms of neuronal loss or damage (see Discussion for a cautionary note about this interpretation). NAA/Cr ratios were assessed for all patients on spectroscopy in both regions. The ratio was particularly low in four patients (patients 2,5,8,19). Of these, patient 5 subsequently died and the remaining three have microcephaly with profound neurological deficit, including motor delay and mental retardation.

DISCUSSION

We have examined 24 patients who were under investigation for suspected metabolic disorder, and compared magnetic resonance spectroscopy observations of brain lactate with measurements of CSF lactate.

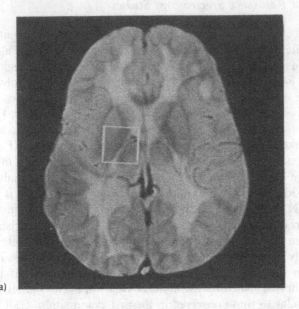

(a)

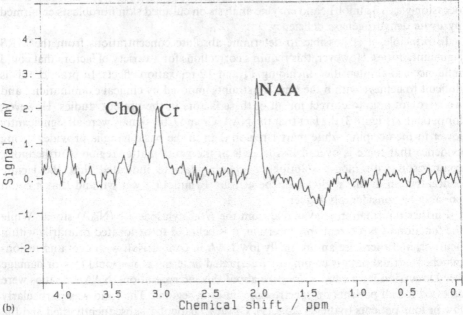

(b)

Figure 3 Magnetic resonance images (a) and (c), and ^{1}H spectra (b) and (d), obtained from a child aged 2 years 10 months (patient 3). The images show generalized white matter change consistent with leukodystrophy. The boxes indicate the positions of the 8 ml cubic volumes of interest from which the spectra were obtained; (b) was from the basal ganglia region, and (d) from abnormal occipital white matter. The dominant contributions to the spectra are from *N*-acetylaspartate (NAA), creatine + phosphocreatine (Cr), choline-containing compounds (Cho), and lactate (Lac).

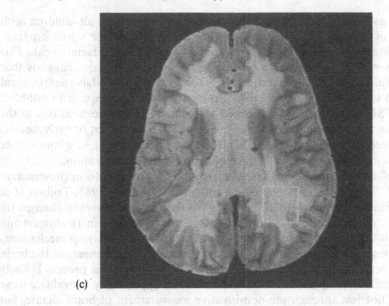

(c)

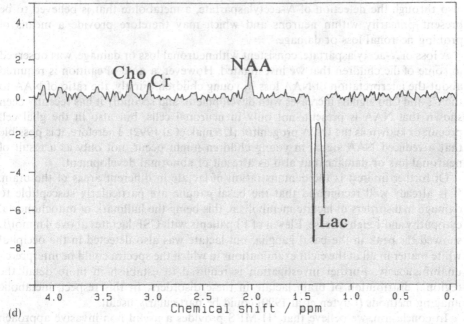

(d)

Figure 3 *continued*

There was good concordance between the two investigations; all children with CSF lactates of 2.5 mmol/L and below had no detectable lactate signal on spectroscopy, while all those with CSF lactates above 4 mmol/L had a lactate peak. This concordance serves to validate both types of measurement, and also suggests that ^{1}H-MRS may have a role in the investigation of those children who have neurological dysfunction in whom screening of blood or urine may not be adequate to establish a diagnosis. CSF lactate is widely assumed to reflect lactate concentrations in the central nervous system but discrepancies are likely with intermittent or early disease. This is illustrated in patient 6 (whose CSF lactate was 3.6 mmol/L), in whom lactate was only demonstrated by MRS at a time of neurological deterioration.

Disorders of lactate metabolism are a heterogeneous group both in presentation and course of illness (Robinson and Sherwood 1984; Robinson 1985; Trijbels et al 1988). A significant number develop neurological dysfunction, showing changes on neuroimaging that correlate with pathological studies at post-mortem (Robinson and Sherwood 1984). However, these are often non-specific and underlying mechanisms responsible remain poorly understood. Because the energy requirements of the brain are largely dependent upon glucose metabolism, any failure in this process is likely to have significant consequences. ^{1}H-MRS provides an opportunity to evaluate these consequences not just through the non-invasive measurement of brain lactate, but also through the detection of N-acetylaspartate, a metabolite that is believed to be present primarily within neurons and which may therefore provide a means of probing neuronal loss or damage.

A loss of N-acetylaspartate, consistent with neuronal loss or damage, was observed in some of the children that we investigated. However, a note of caution is required about the interpretation of NAA loss in young children. Firstly, the ratio of NAA to the Cr and Cho signals increases with development, and secondly it has recently been shown that NAA is present, not only in neuronal cells, but also in the glial cell precursor known as the O-2A progenitor (Urenjak et al 1992). Therefore, it is possible that a reduced NAA signal in young children might occur, not only as a result of neuronal loss or damage, but also as a result of abnormal development.

Of further interest is the demonstration of lactate in different areas of the brain. It is already well recognized that the basal ganglia are particularly susceptible to damage in disorders of lactate metabolism, this being the hallmark of mitochondrial cytopathy and Leigh disease. Eleven of 13 patients with CSF lactates above 4 mmol/L showed the peak in the basal ganglia, but lactate was also detected in the occipital white matter in all of the eight examinations in which the spectra could be interpreted unambiguously. Further investigation is required to establish in more detail the regional distribution of brain lactate in these disorders. In this respect, metabolic imaging methods (Luyten et al 1990) could be particularly useful.

In conclusion, we believe that ^{1}H-MRS provides a useful non-invasive approach for monitoring children with disorders of lactate metabolism. Since MRS and MRI can be performed in the same magnetic resonance examination, MRS observations of brain metabolites, including lactate, can be directly related to structural abnormalities as assessed by MRI. Such integrated magnetic resonance examinations should provide further insights into the mechanisms of brain damage and help to explain the neurological patterns that are observed in these disorders.

ACKNOWLEDGEMENTS

We are grateful for the help of Dr G.K. Brown, Dr K. Bartlett and Mr A. Whitfield for enzyme assays on the patients, Professor B. Lake for histochemistry, and Ms F. Taylor for her advice about assay of lactates.

REFERENCES

Brown GK, Haan EA, Kirby DM, Scholem RD, Wraith JE, Rogers JG, Danks DM (1988) 'Cerebral' lactic acidosis: defects in pyruvate metabolism with profound brain damage and minimal systemic acidosis. *Eur J Pediatr* 147: 10–14.

Bruhn H, Frahm J, Gyngell ML, Merboldt KD, Hanicke W, Sauter R (1989) Cerebral metabolism in man after acute stroke: new observations using localised proton NMR spectroscopy. *Magn Reson Med* 9: 126–131.

Connelly A, Austin SJ, Gadian DG (1991) Localised ^{1}H MRS in the paediatric brain: age and regional dependence. Proc 10th Annual Meeting, Soc Magn Reson Med., Berkeley, CA, USA p. 379.

Detre JA, Wang Z, Bogdan AR, Gusnard DA, Bay CA, Bingham PM and Zimmerman RA (1991) Regional variation in brain lactate in Leigh syndrome by localized ^{1}H magnetic resonance spectroscopy. Ann Neurol 29: 218–221.

Finn JP, Connelly A, Atkinson D (1989) A modified inversion recovery sequence for routine high contrast brain imaging. Proc. 8th Annual Meeting, Soc Magn Reson Med., Berkeley, CA, USA, p. 722.

Frahm J, Bruhn H, Gyngell ML, Merboldt KD, Hanicke W, Sauter R (1989) Localized proton NMR spectroscopy in different regions of the human brain *in vivo*. Relaxation times and concentrations of cerebral metabolites. *Magn Reson Med* 11: 47–63.

Grodd W, Krageloh-Mann I, Klose U, Sauter R (1991) Metabolic and destructive brain disorders in children: findings with localized proton MR spectroscopy. *Radiology* 181: 173–181.

Klose U (1990) *In vivo* proton spectroscopy in presence of eddy currents. *Magn Reson Med* 14: 23–60.

van der Knapp MS, van der Grond J, van Rijen PC, Faber JAJ, Valk J, Willemse K (1990) Age-dependent changes in localized proton and phosphorus MR spectroscopy of the brain. *Radiology* 176: 509–515.

Kollee LAA, Willems JL, De Kort AFM, Monnens LAH, Trijbels JMF (1977) Blood sampling technique for lactate and pyruvate estimation in children. *Ann Clin Biochem* 14: 285–287.

Luyten PR, Marien AJH, Heindel W, van Gerwen PHJ, Herholz K, den Hollander JA, Friedmann G, Heiss W-D (1990) Metabolic imaging of patients with intracranial tumors: H-1 MR spectroscopic imaging and PET. *Radiology* 176: 791–799.

Ordidge RJ, Bendall MR, Gordon RE, Connelly A (1985) Volume selection for *in vivo* spectroscopy. In: Gorvind G, Khatrapal Cl and Saran A, eds. *Magnetic Resonance in Biology and Medicine*. Tata-McGraw-Hill, New Delhi, pp. 387–397.

Robinson BH (1985) The lactic acidaemias. In Lloyd JK, Scriver CR, eds. *Genetic and Metabolic Disease in Paediatrics*. Butterworth, London, pp. 111–139.

Robinson BH, Sherwood WG (1984) Lactic acidaemia. *J Inher Metab Dis* 7 (suppl 1): 69–73.

Shepherd JK, Hall-Craggs MA, Finn JP, Bingham RM (1990) Sedation in children scanned with high field magnetic resonance: the experience at The Hospital for Sick Children, Great Ormond Street. *Br J Radiol* 63: 794–797.

Trijbels JMF, Sengers RCA, Ruitenbeek W, Fischer JC, Bakkeren JAJM, Janssen AJM (1988) Disorders of the mitochondrial respiratory chain: clinical manifestations and diagnostic approach. *Eur J Pediatr* 148: 92–97.

Urenjak J, Williams SR, Gadian DG, Noble M (1992) Specific expression of *N*-acetyl-aspartate in neurons, oligodendrocyte-type 2 astrocye (O-2A) progenitors and immature oligodendrocytes *in vivo*. *J Neurochem* 59: 55–61.

Van Erven PMM, Gabreels FJM, Ruitenbeek W, Den Hartog MR, Fischer JC, Renier WO, Trijbels JMF, Sloff JL, Janssen AJM (1985) Subacute necrotizing encephalomyelopathy (Leigh syndrome) associated with disturbed oxidation of pyruvate, malate and 2-oxoglutarate in muscle and liver. *Acta Neurol Scand* 72: 36–42.

Journal of Inherited Metabolic Disease

Aims and Scope of the Journal

The *Journal of Inherited Metabolic Disease* is the official scientific and clinical journal of the Society for the Study of Inborn Errors of Metabolism. The aim of this international and mutidisciplinary journal is to provide otherwise unavailable information on inherited metabolic disorders covering clinical (medical, dental and veterinary), biochemical (including molecular genetics), genetic, experimental (including cell biology), theoretical, epidemiological, ethical and counselling aspects. Widespread and efficient communication between professional workers should improve the handling and understanding of inherited disorders.

The journal publishes papers, case and short reports, short communications, invited articles which are generally reviews, and book reviews, *Papers for submission* and correspondence should be written in English and sent to:

> The Editor-in-Chief
> Journal of Inherited Metabolic Disease
> Kluwer Academic Publishers
> PO Box 55, Lancaster, LA1 1PE, United Kingdom.

Acceptance of articles for publication is at the discretion of the editors. Authors are advised to consult a current issues of the journal before submission. It is a condition of acceptance that all articles have been and will not be published elsewhere in substantially the same form. The submitting author must have circulated the article to and have secured agreement from all co-authors before submission of the article. The absence of previous similar or simultaneous publications, their inspection of the manuscript, their substantial contribution to the work and their agreement to submission must be confirmed by all authors in a signed letter or letters on submission. It should be noted that the conditions are later confirmed by the corresponding author in a copyright transfer form when the paper is accepted.

Detailed *Instructions to Authors* is published in issue no.1 of each volume of the journal, and in other issues as space permits, and is listed in the Contents.

Ethical considerations

The editors reserve the right to reject for publication work which, even though scientifically sound, they consider should not have been undertaken on ethical grounds. Where there is a difference of opinion, authors may be asked to explain in their publication why the work was undertaken. There are no absolute ethical standards, and studies on children present special problems. In this last respect the editors agree with the policy stated by the *Archives of Disease in Childhood* 42 (1967) 109; 48 (1973) 751–752; 54 (1978) 441–442:

> The purely legal aspect of studies in children are discussed in the *Lancet* 2 (1977) 754–755; *Archives of Disease in Childhood* 53 (1978) 443–446

Books for Review should be sent to The Editor-in-Chief at the above address.

Offprints. 25 offprints are provided free of charge. Authors will receive a form for ordering an extra 125 offprints with the proofs of their paper.

No page charges are levied on authors or their instructions.

Advertisements: As well as commerical advertisements, announcements of forthcoming scientific meetings and other material relevant to the journal can be included, at rates available from the publishers. All advertising is subject to the discretion of the editors.

Consent to publish in this journal entails the author's irrevocable and exclusive authorisation of the publisher to collect on behalf of the copyright owners, SSIEM and the publishers, any sums or considerations for copying or reproduction payable by third parties (as mentioned in article 17, paragraph 2, of the Dutch Copyright act of 1912 and in the Royal Decree of June 20, 1974 (S.351) pursuant to article 16b of the Dutch Copyright act of 1912) and/or to act in or out of court in connection therewith.

Publication programme, 1993: Volume 16 (6 issues)
Subscriptions should be sent to: **Kluwer Academic Publishers Group, PO Box 322, 3300 AH Dordrecht, The Netherlands,** or at **PO Box 358, Accord Station, Hingham, MA 02018-0358, USA,** or to any subscription agent.
Changes of mailing address should be notified together with your latest label.
Subscription prices, per volume (6 issues): Dfl 490.- plus postage: Dfl 46.- (Dfl 536.- per annum).
Subscription prices, per volume (bimonthly): US$335.00.
Second Class Postage paid at Rahway, NJ. USPS No. 757-750.
US Mailing Agent: Expediters of the Printed Word Inc., 2323 Randolph Ave., Avenel, NJ 07001, USA.
Published by Kluwer Academic Publishers, PO Box 55, Lancaster, LA1 1PE, UK.
Postmaster: please send all address corrections to: *Journal of Inherited Metabolic Disease,*
c/o Expediters of the Printed Word Inc., 2323 Randolph Ave, Avenel, NJ 07001, USA.

Contents (continued from outside back cover)

ORDER FORM

The Review Issues of the journal containing papers presented at the Annual Meetings of the SSIEM are also available separately, price Dfl. 130.- or US$ 78.50 each.

You can order either from your usual bookshop or, in case of difficulty, from:
United Kingdom and the Rest of the World: Kluwer Academic Publishers, Order Department, PO Box 322, 3300 AH Dordrecht, The Netherlands
USA/Canada: Kluwer Academic Publishers, Order Dept.-M, PO Box 358, Accord Station, Hingham, MA 02061-0358, USA

Please send me:

Inherited Metabolic Diseases and the Brain; 1993; Guest Editor J. Jaeken:
ISBN 0-7923-8837-2; Vol.16 No.4; 30th Annual Meeting   . . . . . copy(s)

Mitochondrial DNA and Associated Disorders; The X Chromosome; 1992; Guest
Editor D.P. Brenton: ISBN 0-7923-8800-3; Vol.15 No.4; 29th Annual Meeting   . . . . . copy(s)

The Liver and Inherited Metabolic Disorders; 1991; Guest Editor A. Green:
ISBN 0-7923-8982-4; Vol.14 No.4; 28th Annual Meeting   . . . . . copy(s)

Carbohydrate and Glycoprotein Metabolism; Maternal Phenylketonuria; 1990;
Guest Editor W. Endres: ISBN 0-7923-8947-6; Vol.13 No.4; 27th Annual Meeting . . . . . copy(s)

I enclose a cheque for . . . . (made payable to "Kluwer Academic Publishers")

OR I authorise you to charge my credit account:

Card: . . . . Number: . . . .

Expiry date: . . . . Signature: . . . .

OR Please invoice me (a small charge will be made for postage and packing)

Name [please print] . . . .

Address [please print] . . . . , ...

. . . .

. . . .

ORDER FORM